JAGADISH CHANDRA BOSE was born on November 30, 1858, Mymensingh, Bengal, India (now in Bangladesh), and died on November 23, 1937, Giridih, Bihar.

His work as both a physiologist and physicist led to the invention of highly sensitive instruments for the detection of minute responses by living organisms to external stimuli. This enabled him to measure the similarities in response between animal and plant tissues noted by many later researchers.

Bose's experiments on the quasi-optical properties of very short radio waves led him to make improvements on the *coherer*, an early form of radio detector, which contributed to the development of solid-state physics.

After earning a degree from the University of Cambridge (1884), Bose served as professor of physical science (1885–1915) at Presidency College, Calcutta. In 1917, he founded the Bose Institute in Calcutta, which still exists today (*www.jcbose.ac.in*).

To facilitate his research, he constructed automatic recorders capable of registering extremely slight movements; these instruments produced some striking results, such as his demonstration of the sense of feeling in plants.

Bose also found that non-living matter exhibits the same types of response to stimuli as do both animal and plant matter. This demonstration that everything exists in the field of consciousness was one of his most important discoveries.

# RESPONSE IN THE LIVING AND NON-LIVING

Jagadish Chandra Bose

With an introduction by
Paramahansa Yogananda

A DISTANT MIRROR

ISBN 978-1548845810

First published 1902.
This edition 2017.

Available in paperback, Kindle and epub.

**A Distant Mirror**
*adistantmirror.com.au*
*admin@adistantmirror.com.au*

# PREFACE

I have in this present work put in a connected and a more complete form results, some of which have been published in the following papers:

*De la Généralité des Phénomènes Moléculaires produits par l'Electricité sur la matière Inorganique et sur la matière Vivante.* (Travaux du Congrès International de Physique. Paris, 1900.)

*On the Similarity of Effect of Electrical Stimulus on Inorganic and Living Substances.* (Report, Bradford Meeting British Association, 1900. – Electrician.)

*Response of Inorganic Matter to Stimulus.* (Friday evening discourse, Royal Institution, May 1901.)

*On Electric Response of Inorganic Substances. Preliminary Notice.* (Royal Society, June 1901.)

*On Electric Response of Ordinary Plants under Mechanical Stimulus.* (Journal Linnean Society, 1902.)

*Sur la Réponse Electrique dans les Métaux, les Tissus Animaux et Végétaux.* (Société de Physique, Paris, 1902.)

*On the Electro-Motive Wave accompanying Mechanical Disturbance in Metals in contact with Electrolyte.* (Proceedings Royal Society, vol. 70.)

*On the Strain Theory of Vision and of Photographic Action.* (Journal Royal Photographic Society, vol. xxvi.)

These investigations were commenced in India, and I take this opportunity to express my grateful acknowledgments to the managers of the Royal Institution, for the facilities offered me to complete them at the Davy-Faraday Laboratory.

J. C. Bose.
Davy-Faraday Laboratory, Royal Institution,
London, May 1902.

# CONTENTS

**Introduction / 15**

**1** The mechanical response of living substances / 27
Mechanical response – Different kinds of stimuli
– Myograph – Characteristics of response curve: period,
amplitude, form – Modification of response curves

**2** Electric response / 31
Conditions for obtaining electric response – Method
of injury – Current of injury – Injured end, cuproid:
uninjured, zincoid – Current of response in nerve from
more excited to less excited – Difficulties of present
nomenclature – Electric recorder – Two types of response,
positive and negative – Universal applicability of electric
mode of response – Electric response a measure of
physiological activity – Electric response in plants

**3** Electric response in plants –
method of negative variation / 44
Negative variation – Response recorder – Photographic
recorder – Compensator – Means of graduating intensity
of stimulus – Spring tapper and torsional vibrator
– Intensity of stimulus dependent on amplitude of vibration
– Effectiveness of stimulus dependent on rapidity also

**4** Electric response in plants – block method / 55
Method of block – Advantages of block method – Plant
response a physiological phenomenon – Abolition of
response by anæsthetics and poisons – Abolition of response
when plant is killed by hot water

**5** Plant response – on the effects of
single stimulus and of superposed stimuli / 63
Effect of single stimulus – Superposition of stimuli
– Additive effect – Staircase effect – Fatigue – No fatigue
when sufficient interval between stimuli – Apparent fatigue
when stimulation frequency is increased – Fatigue under
continuous stimulation

**6** Plant response – on diphasic variation / 73

Diphasic variation – Positive after-effect and positive response – Radial E.M. variation

**7** Plant response – on the relation between stimulus and response / 79

Increased response with increasing stimulus – Apparent diminution of response with excessively strong stimulus

**8** Plant response – on the influence of temperature / 89

Effect of very low temperature – Influence of high temperature – Determination of death-point – Increased response as after-effect of temperature variation – Death of plant and abolition of response by the action of steam

**9** Plant response – effect of anæsthetics and poisons / 100

Effect of anæsthetics, a test of vital character of response – Effect of chloroform – Effect of chloral – Effect of formalin – Method in which response is unaffected by variation of resistance – Advantage of block method – Effect of dose

**10** Response in metals / 108

Is response found in inorganic substances? – Experiment on tin, block method – Anomalies of existing terminology – Response by method of depression – Response by method of exaltation

**11** Inorganic response – modified apparatus to exhibit response in metals / 119

Conditions of obtaining quantitative measurements – Modification of the block method – Vibration cell – Application of stimulus – Graduation of the intensity of stimulus – Considerations showing that electric response is due to molecular disturbance – Test experiment – Molecular voltaic cell

**12** Inorganic response – method of ensuring consistent results / 127

Preparation of wire – Effect of single stimulus

**13** Inorganic response – molecular mobility: its influence on response / 131

Effects of molecular inertia – Prolongation of period of recovery by overstrain – Molecular model – Reduction of molecular sluggishness attended by quickened recovery and heightened response – Effect of temperature – Modification of latent period and period of recovery by the action of chemical reagents – Diphasic variation

**14** Inorganic response – fatigue, staircase, and modified response / 144

Fatigue in metals – Fatigue under continuous stimulation – Staircase effect – Reversed responses due to molecular modification in nerve and in metal, and their transformation into normal after continuous stimulation – Increased response after continuous stimulation

**15** Inorganic response – relation between stimulus and response – superposition of stimuli / 157

Relation between stimulus and response – Magnetic analogue – Increase of response with increasing stimulus – Threshold of response – Superposition of stimuli – Hysteresis

**16** Inorganic response – effect of chemical reagents / 165

Action of chemical reagents – Action of stimulants on metals – Action of depressants on metals – Effect of poisons on metals – Opposite effect of large and small doses

**17** On the stimulus of light and retinal currents / 174

Visual impulse: (1) chemical theory; (2) electrical theory – Retinal currents – Normal response positive – Inorganic response under stimulus of light – Typical experiment on the electrical effect induced by light

**18** Inorganic response – influence of various conditions on the response to stimulus of light / 184

Effect of temperature – Effect of increasing length of exposure – Relation between intensity of light and magnitude of response – After-oscillation – Abnormal effects: (1) preliminary negative twitch; (2) reversal of response; (3) transient positive twitch on cessation of light; (4) decline and reversal – Résumé

**19** Visual analogues / 195

Effect of light of short duration – After-oscillation – Positive and negative after-images – Binocular alternation of vision – Period of alternation modified by physical condition – After-images and their revival – Unconscious visual impression.

**20** General Survey and Conclusion / 205

# ILLUSTRATIONS

**Figure**

1. Mechanical lever recorder - 29
2. Electric method of detecting nerve response - 32
3. Diagram showing injured end of nerve corresponds to copper in a voltaic element - 34
4. Electric recorder - 37
5. Simultaneous record of mechanical and electrical responses - 39
6. Negative variation in plants - 45
7. Photographic record of negative variation in plants - 46
8. Response recorder - 48
9. The compensator - 49
10. The spring tapper - 50
11. The torsional vibrator - 51
12. Response in plant to mechanical tap or vibration - 52
13. Influence of suddenness on the efficiency of stimulus - 53
14. The method of block - 56
15. Response in plant completely immersed under water - 57
16. Uniform responses in plant - 63
17. Fusion of effect under rapidly succeeding stimuli in muscle and in plant - 64
18. Additive effect of singly ineffective stimuli on plant - 65
19. 'Staircase effect' in plant - 66
20. Appearance of fatigue in plant under shortened period of rest - 68
21. Fatigue in celery - 69
22. Fatigue in cauliflower stalk - 69
23. Fatigue from previous overstrain - 70
24. Fatigue under continuous stimulation in muscle and celery - 71
25. Effect of rest in removal of fatigue in plant - 72
26. Diphasic variation in plant - 75
27. Abnormal positive responses in stale turnip leaf stalk transformed into normal negative under strong stimulation - 76

28. Abnormal positive passing into normal negative in a stale specimen of caulifower - 77
29. Radial E.M. Variation - 78
30. Curves showing the relation between intensity of stimulus and response in muscle and nerve - 80
31. Increasing responses to increasing stimuli (taps) in plants - 80
32. Increasing responses to increasing vibrational stimuli in plants - 81
33. Responses to increasing stimuli in fresh and stale specimens of plants - 82
34. Apparent diminution of response caused by fatigue under strong stimulation - 85
35. Diminution of response in eucharis lily at low temperature - 89
36. Records showing the difference in the effects of low temperature on Ivy, Holly, and Eucharis Lily - 90
37. Plant chamber for studying the effect of temperature and anæsthetics - 92
38. Effect of high temperature on plant response - 93
39. After-effect on the response due to temperature variation - 94
40. Records of responses in eucharis lily during rise and fall of temperature - 96
41. Curve showing variation of sensitiveness during a cycle of temperature variation - 96
42. Record of effect of steam in abolition of response at death of plant - 98
43. Effect of chloroform on nerve response - 101
44. Effect of chloroform on the responses of carrot - 103
45. Action of chloral hydrate on plant responses - 103
46. Action of formalin on radish - 104
47. Action of sodium hydrate in abolishing the response in plant - 106
48. Stimulating action of poison in small doses in plants - 107
49. The poisonous effect of stronger dose of KOH - 107
50. Block method for obtaining response in tin - 109
51. Response to mechanical stimulation in a Zn-Cu couple - 112
52. Electric response in metal by the method of relative depression (negative variation) - 115
53. Method of relative exaltation - 116
54. Various cases of positive and negative variation - 117

55. Modifications of the block method for exhibiting electric response in metals - 121

56. Equal and opposite responses given by two ends of the wire - 123

57. Top view of the vibration cell - 124

58. Influence of annealing in the enhancement of response in metals - 128

59. Uniform electric responses in metals - 129

60. Persistence of after-effect - 132

61. Prolongation of period of recovery after overstrain - 133

62. Model showing the effect of friction - 134

63, 64. Effects of removal of molecular sluggishness in quickened recovery and heightened response in metals - 136, 137

65. Effect of temperature on response in metals - 138

66. Diphasic variation in metals - 140

67. Negative, diphasic, and positive resultant response in metals - 141

68. Continuous transformation from negative to positive through intermediate diphasic response - 142

69. Fatigue in muscle - 144

70. Fatigue in platinum - 145

71. Fatigue in tin - 145

72. Appearance of fatigue due to shortening the period of recovery - 146

73. Fatigue in metal under continuous stimulation - 147

74. 'Staircase' response in muscle and in metal - 148

75. Abnormal response in nerve converted into normal under continued stimulation - 150

76, 77. Abnormal response in tin and platinum converted into normal under continued stimulation - 151

78. Gradual transition from abnormal to normal response in platinum - 151

79. Increase of response in nerve after continuous stimulation - 152

80, 81. Response in tin and platinum enhanced after continuous stimulation - 153

82. Magnetic analogue - 158

83. Records of responses to increasing stimuli in tin - 160

84. A second set of records with a different specimen of tin - 161

85. Ineffective stimulus becoming effective by superposition - 161

86. Incomplete and complete fusion of effects - 163

87. Cyclic curve for maximum effects showing hysteresis - 164
88. Action of poison in abolishing response in nerve - 165
89. Action of stimulant on tin - 167
90. Action of stimulant on platinum - 168
91. Depressing effect of KBr on tin - 168
92. Abolition of response in metals by 'poison' - 169
93. 'Molecular arrest' by the action of 'poison' - 171
94. Opposite effects of small and large doses on the response in metals - 172
95. Retinal response to light - 176
96. Response of sensitive cell to light - 178
97. Typical experiment on the E.M. variation produced by light - 180
98. Modification of the photo-sensitive cell - 181
99. Responses in frog's retina - 183
100. Responses in sensitive photo-cell - 180
101. Effect of temperature on the response to light stimulus - 185
102. Effect of duration of exposure on the response - 185
103. Responses of sensitive cell to increasing intensities of light - 187
104. Relation between the intensity of light and magnitude of response - 188
105. After-oscillation - 189
106. Transient positive increase of response in the frog's retina on the cessation of light - 190
107. Transient positive increase of response in the sensitive cell - 190
108. Decline under the continuous action of light - 192
109. Certain after-effects of light - 194
110. After-effect of light of short duration - 196
111. Stereoscopic design for the exhibition of binocular alternation of vision - 200
112. Uniform responses in nerve, plant, and metal - 207
113. Fatigue in muscle, plant, and metal - 208
114. 'Staircase' effect in muscle, plant, and metal - 209
115. Increase of response after continuous stimulation in nerve and metal - 210
116. Modified abnormal response in nerve and metal transformed into normal response after continuous stimulation - 211
117. Action of the same 'poison' in the abolition of response in nerve, plant, and metal - 213

*Paramahansa Yogananda*

# INTRODUCTION

The following is from Paramahansa Yogananda's *Autobiography of a Yogi*. It describes conversations between the swami and Jagadish Bose.

---

## *India's Great Scientist, J. C. Bose*

"Jagadish Chandra Bose's wireless inventions antedated those of Marconi."

Overhearing this provocative remark, I walked closer to a sidewalk group of professors engaged in scientific discussion. If my motive in joining them was racial pride, I regret it. I cannot deny my keen interest in evidence that India can play a leading part in physics, and not metaphysics alone.

"What do you mean, sir?"

The professor obligingly explained. "Bose was the first one to invent a wireless coherer and an instrument for indicating the refraction of electric waves. But the Indian scientist did not exploit his inventions commercially. He soon turned his attention from the inorganic to the organic world. His revolutionary discoveries as a plant physiologist are outpacing even his radical achievements as a physicist."

I politely thanked my mentor. He added, "The great scientist is one of my brother professors at Presidency College."

\*

I paid a visit the next day to the sage at his home, which was close to mine on Gurpar Road. I had long admired him from a respectful distance. The grave and retiring botanist greeted me graciously. He was a handsome, robust man in his fifties, with thick hair, broad forehead, and the abstracted eyes of a dreamer. The precision in his tones revealed the lifelong scientific habit.

"I have recently returned from an expedition to scientific societies of the West. Their members exhibited intense interest in delicate instruments of my invention which demonstrate the indivisible unity of all life.[1] The Bose crescograph has the enormity of ten million magnifications. The microscope enlarges only a few thousand times; yet it brought vital impetus to biological science. The crescograph opens incalculable vistas."

"You have done much, sir, to hasten the embrace of East and West in the impersonal arms of science."

"I was educated at Cambridge. How admirable is the Western method of submitting all theory to scrupulous experimental verification! That empirical procedure has gone hand in hand with the gift for introspection which is my Eastern heritage. Together they have enabled me to sunder the silences of natural realms long uncommunicative. The telltale charts of my crescograph[2] are evidence for the most skeptical that plants have a sensitive nervous system and a varied emotional life. Love, hate, joy, fear, pleasure, pain, excitability, stupor, and countless appropriate responses to stimuli are as universal in plants as in animals."

"The unique throb of life in all creation could seem only poetic imagery before your advent, Professor! A saint I once knew would never pluck flowers. 'Shall I rob the rosebush of its pride in beauty? Shall I cruelly affront its dignity by my rude divestment?' His sympathetic words are verified literally through your discoveries!"

"The poet is intimate with truth, while the scientist approaches awkwardly. Come someday to my laboratory and see the unequivocable testimony of the crescograph."

Gratefully I accepted the invitation, and took my departure. I heard later that the botanist had left Presidency College, and was planning a research center in Calcutta.

When the Bose Institute was opened, I attended the dedicatory services. Enthusiastic hundreds strolled over the premises. I was charmed with the artistry and spiritual symbolism of the new home of science. Its front gate, I noted, was a centuried relic from a distant shrine. Behind the lotus[3] fountain, a sculptured female figure with a torch conveyed the Indian respect for woman as the immortal light-bearer. The garden held a small temple consecrated to the Noumenon beyond phenomena. Thought of the divine incorporeity was suggested by absence of any altar-image.

Bose's speech on this great occasion might have issued from the lips of one of the inspired ancient rishis.

"I dedicate today this Institute as not merely a laboratory but a temple." His reverent solemnity stole like an unseen cloak over the crowded auditorium. "In the pursuit of my investigations I was unconsciously led into the border region of physics and physiology. To my amazement, I found boundary lines vanishing, and points of contact emerging, between the realms of the living and the non-living. Inorganic matter was perceived as anything but inert; it was athrill under the action of multitudinous forces.

"A universal reaction seemed to bring metal, plant and animal under a common law. They all exhibited essentially the same phenomena of fatigue and depression, with possibilities of recovery and of exaltation, as well as the permanent irresponsiveness associated with death. Filled with awe at this stupendous generalization, it was with great hope that I announced my results before the Royal Society—results demonstrated by experiments. But the physiologists present advised me to confine myself to physical investigations, in which my success had been assured, rather than encroach on their preserves. I had unwittingly strayed into the domain of an unfamiliar caste system and so offended its etiquette.

"An unconscious theological bias was also present, which confounds ignorance with faith. It is often forgotten that He who surrounded us with this ever-evolving mystery of creation has also implanted in us the desire to question and understand. Through many years of miscomprehension, I came to know that the life of a devotee of science is inevitably filled with unending struggle. It is for him to cast his life as an ardent offering—regarding gain and loss, success and failure, as one.

"In time the leading scientific societies of the world accepted my theories and results, and recognized the importance of the Indian contribution to science.[4] Can anything small or circumscribed ever satisfy the mind of India? By a continuous living tradition, and a vital power of rejuvenescence, this land has readjusted itself through unnumbered transformations. Indians have always arisen who, discarding the immediate and absorbing prize of the hour, have sought for the realization of the highest ideals in life—not through passive renunciation, but through active struggle. The weakling who has refused the conflict, acquiring nothing, has had nothing to renounce. He alone who has striven and won can enrich the world by bestowing the fruits of his victorious experience.

"The work already carried out in the Bose laboratory on the response of matter, and the unexpected revelations in plant life, have opened out very extended regions of inquiry in physics, in physiology, in medicine, in agriculture, and even in psychology. Problems hitherto regarded as insoluble have now been brought within the sphere of experimental investigation.

"But high success is not to be obtained without rigid exactitude. Hence the long battery of super-sensitive instruments and apparatus of my design, which stand before you today in their cases in the entrance hall. They tell you of the protracted efforts to get behind the deceptive seeming into the reality that remains unseen, of the continuous toil and persistence and resourcefulness called forth to overcome human limitations. All creative scientists know that the true laboratory is the mind, where behind illusions they uncover the laws of truth.

"The lectures given here will not be mere repetitions of second-hand knowledge. They will announce new discoveries, demonstrated for the first time in these halls. Through regular publication of the work of the Institute, these Indian contributions will reach the whole world. They will become public property. No patents will ever be taken. The spirit of our national culture demands that we should forever be free from the desecration of utilizing knowledge only for personal gain.

"It is my further wish that the facilities of this Institute be available, so far as possible, to workers from all countries. In this I am attempting to carry on the traditions of my country. So far back as twenty-five centuries, India welcomed to its ancient universities, at Nalanda and Taxila, scholars from all parts of the world.

"Although science is neither of the East nor of the West but rather international in its universality, yet India is specially fitted to make great contributions.[5] The burning Indian imagination, which can extort new order out of a mass of apparently contradictory facts, is held in check by the habit of concentration. This restraint confers the power to hold the mind to the pursuit of truth with an infinite patience."

Tears stood in my eyes at the scientist's concluding words. Is "patience" not indeed a synonym of India, confounding Time and the historians alike?

\*

I visited the research center again, soon after the day of opening. The great botanist, mindful of his promise, took me to his quiet laboratory.

"I will attach the crescograph to this fern; the magnification is tremendous. If a snail's crawl were enlarged in the same proportion, the creature would appear to be traveling like an express train!"

My gaze was fixed eagerly on the screen which reflected the magnified fern-shadow. Minute life-movements were now clearly perceptible; the plant was growing very slowly before my fascinated eyes. The scientist touched the tip of the fern with a small metal bar. The developing pantomime came to an abrupt halt, resuming the eloquent rhythms as soon as the rod was withdrawn.

"You saw how any slight outside interference is detrimental to the sensitive tissues," Bose remarked. "Watch; I will now administer chloroform, and then give an antidote."

The effect of the chloroform discontinued all growth; the antidote was revivifying. The evolutionary gestures on the screen held me more raptly than a "movie" plot. My companion (here in the role of villain) thrust a sharp instrument through a part of the fern; pain was indicated by spasmodic flutters. When he passed a razor partially through the stem, the shadow was violently agitated, then stilled itself with the final punctuation of death.

"By first chloroforming a huge tree, I achieved a successful transplantation. Usually, such monarchs of the forest die very quickly after being moved." Jagadis smiled happily as he recounted the life-saving maneuver. "Graphs of my delicate apparatus have proved that trees possess a circulatory system; their sap movements correspond to the blood pressure of animal bodies. The ascent of sap is not explicable on the mechanical grounds ordinarily advanced, such as capillary attraction. The phenomenon has been solved through the crescograph as the activity of living cells. Peristaltic waves issue from a cylindrical tube which extends down a tree and serves as an actual heart! The more deeply we perceive, the more striking becomes the evidence that a uniform plan links every form in manifold nature."

The great scientist pointed to another Bose instrument.

"I will show you experiments on a piece of tin. The life-force in metals responds adversely or beneficially to stimuli. Ink markings will register the various reactions."

Deeply engrossed, I watched the graph which recorded the characteristic waves of atomic structure. When the professor applied chloroform to the tin, the vibratory writings stopped. They recommenced as the metal slowly regained its normal state. My companion dispensed a poisonous chemical. Simultaneous with the quivering end of the tin, the needle dramatically wrote on the chart a death-notice.

"Bose instruments have demonstrated that metals, such as the steel used in scissors and machinery, are subject to fatigue, and regain efficiency by periodic rest. The life-pulse in metals is seriously harmed or even extinguished through the application of electric currents or heavy pressure."

I looked around the room at the numerous inventions, eloquent testimony of a tireless ingenuity.

"Sir, it is lamentable that mass agricultural development is not speeded by fuller use of your marvelous mechanisms. Would it not be easily possible to employ some of them in quick laboratory experiments to indicate the influence of various types of fertilizers on plant growth?"

"You are right. Countless uses of Bose instruments will be made by future generations. The scientist seldom knows contemporaneous reward; it is enough to possess the joy of creative service."

With expressions of unreserved gratitude to the indefatigable sage, I took my leave. "Can the astonishing fertility of his genius ever be exhausted?" I thought.

No diminution came with the years. Inventing an intricate instrument, the "Resonant Cardiograph," Bose then pursued extensive researches on innumerable Indian plants. An enormous unsuspected pharmacopoeia of useful drugs was revealed. The cardiograph is constructed with an unerring accuracy by which a one-hundredth part of a second is indicated on a graph. Resonant records measure infinitesimal pulsations in plant, animal and human structure. The great botanist predicted that use of his cardiograph will lead to vivisection on plants instead of animals.

"Side by side recordings of the effects of a medicine given simultaneously to a plant and an animal have shown astounding unanimity in result," he pointed out. "Everything in man has been foreshadowed in the plant. Experimentation on vegetation will contribute to lessening of human suffering."

\*

Years later Bose's pioneer plant findings were substantiated by other scientists. Work done in 1938 at Columbia University was reported by The New York Times as follows:

It has been determined within the past few years that when the nerves transmit messages between the brain and other parts of the body, tiny electrical impulses are being generated. These impulses have been measured by delicate galvanometers and magnified millions of times by modern amplifying apparatus. Until now no satisfactory method had been found to study the passages of the impulses along the nerve fibers in living animals or man because of the great speed with which these impulses travel.

Drs. K. S. Cole and H. J. Curtis reported having discovered that the long single cells of the fresh-water plant nitella, used frequently in goldfish bowls, are virtually identical with those of single nerve fibers. Furthermore, they found that nitella fibers, on being excited, propagate electrical waves that are similar in every way, except velocity, to those of the nerve fibers in animals and man. The electrical nerve impulses in the plant were found to be much slower than those in animals. This discovery was therefore seized upon by the Columbia workers as a means for taking slow motion pictures of the passage of the electrical impulses in nerves.

The nitella plant thus may become a sort of Rosetta stone for deciphering the closely guarded secrets close to the very borderland of mind and matter.

The poet Rabindranath Tagore was a stalwart friend of India's idealistic scientist. To him, the sweet Bengali singer addressed the following lines:[6]

> O Hermit, call thou in the authentic words
> Of that old hymn called Sama; "Rise! Awake!"
> Call to the man who boasts his shastric lore
> From vain pedantic wranglings profitless,
> Call to that foolish braggart to come forth
> Out on the face of nature, this broad earth,
> Send forth this call unto thy scholar band;
> Together round thy sacrifice of fire
> Let them all gather. So may our India,
> Our ancient land unto herself return
> O once again return to steadfast work,
> To duty and devotion, to her trance
> Of earnest meditation; let her sit
> Once more unruffled, greedless, strifeless, pure,
> O once again upon her lofty seat
> And platform, teacher of all lands.

\* \* \*

### Notes

1. "All science is transcendental or else passes away. Botany is now acquiring the right theory—the avatars of Brahma will presently be the textbooks of natural history."—Emerson.

2. From the Latin root, *crescere*, to increase. For his crescograph and other inventions, Bose was knighted in 1917.

3. The lotus flower is an ancient divine symbol in India; its unfolding petals suggest the expansion of the soul; the growth of its pure beauty from the mud of its origin holds a benign spiritual promise.

4. "At present, only the sheerest accident brings India into the purview of the American college student. Eight universities (Harvard, Yale, Columbia, Princeton, Johns Hopkins, Pennsylvania, Chicago, and California) have chairs of Indology or Sanskrit, but India is virtually unrepresented in departments of history, philosophy, fine arts, political science, sociology, or any of the other departments of intellectual experience in which, as we have seen, India has made great contributions. . . . We believe, consequently, that no department of study, particularly in the humanities, in any major university can be fully equipped without a properly trained specialist in the Indic phases of its discipline. We believe, too, that every college which aims to prepare its graduates for intelligent work in the world which is to be theirs to live in, must have on its staff a scholar competent in the civilization of India."

   —Extracts from an article by Professor W. Norman Brown of the University of Pennsylvania which appeared in the May, 1939, issue of the *Bulletin of the American Council of Learned Societies*, 907 15th St., Washington, D. C. This issue (#28) contains over 100 pages of a bibliography for Indic Studies.

5. The atomic structure of matter was well-known to the ancient Hindus. One of the six systems of Indian philosophy is Vaisesika, from the Sanskrit root *visesas*, "atomic individuality." One of the foremost Vaisesika expounders was Aulukya, also called Kanada, "the atom-eater," born about 2800 years ago. In an article in East-West, April, 1934, a summary of Vaisesika scientific knowledge was given as follows:

   "Though the modern 'atomic theory' is generally considered a new advance of science, it was brilliantly expounded long ago by Kanada, 'the atom-eater.' The Sanskrit *anu* can be properly translated as 'atom' in the latter's literal Greek sense of 'uncut' or indivisible. Other scientific expositions of Vaisesika treatises include:

(1) the movement of needles toward magnets,

(2) the circulation of water in plants,

(3) akash or ether, inert and structureless, as a basis for transmitting subtle forces,

(4) the solar fire as the cause of all other forms of heat,

(5) heat as the cause of molecular change,

(6) the law of gravitation as caused by the quality that inheres in earth-atoms to give them their attractive power or downward pull,

(7) the kinetic nature of all energy; causation as always rooted in an expenditure of energy or a redistribution of motion,

(8) universal dissolution through the disintegration of atoms,

(9) the radiation of heat and light rays, infinitely small particles, darting forth in all directions with inconceivable speed (the modern 'cosmic rays' theory),

(10) the relativity of time and space.

Vaisesika assigned the origin of the world to atoms, eternal in their nature, i.e., their ultimate peculiarities. These atoms were regarded as possessing an incessant vibratory motion. The recent discovery that an atom is a miniature solar system would be no news to the old Vaisesika philosophers, who also reduced time to its furthest mathematical concept by describing the smallest unit of time (kala) as the period taken by an atom to traverse its own unit of space."

6. Translated from the Bengali of Rabindranath Tagore, by Manmohan Ghosh, in Viswa-Bharati.

*Jagadish Bose*

# Chapter I
# THE MECHANICAL RESPONSE OF LIVING SUBSTANCES

*Mechanical response – Different kinds of stimuli – Myograph – Characteristics of the response curve: period, amplitude, form – Modification of response curves.*

One of the most striking effects of external disturbance on certain types of living substance is a visible change of form. Thus, a piece of muscle when pinched contracts. The external disturbance which produced this change is called the stimulus. The body which is thus capable of responding is said to be irritable or excitable. A stimulus thus produces a state of excitability which may sometimes be expressed by change of form.

## Mechanical response to different kinds of stimuli

This reaction under stimulus is seen even in the lowest organisms; in some of the amœboid rhizopods, for instance. These lumpy protoplasmic bodies, usually elongated while creeping, if mechanically jarred, contract into a spherical form.

If, instead of mechanical disturbance, we apply salt solution, they again contract, in the same way as before. Similar effects are produced by sudden illumination, or by rise of temperature, or by electric shock.

A living substance may thus be put into an excitatory state by either mechanical, chemical, thermal, electrical, or light stimulus. Not only does the point stimulated show the effect of stimulus, but that effect may sometimes be conducted even to a considerable distance.

This power of conducting stimulus, though common to all living substances, is present in very different degrees. While in some forms of animal tissue irritation spreads, at a very slow rate, only to points in close neighbourhood, in other forms, as for example in nerves, conduction is very rapid and reaches far.

The visible mode of response by change of form may perhaps be best studied in a piece of muscle. When this is pinched, or an electrical shock is sent through it, it becomes shorter and broader. A responsive twitch is thus produced. The excitatory state then disappears, and the muscle is seen to relax into its normal form.

## Mechanical lever recorder

In the case of contraction of muscle, the effect is very quick, the twitch takes place in too short a time for detailed observation by ordinary means.

A myographic apparatus is therefore used, by means of which the changes in the muscle are self-recorded. Thus we obtain a history of its change and recovery from the change.

The muscle is connected to one end of a writing lever. When the muscle contracts, the tracing point is pulled up in one direction, say to the right. The extent of this pull depends on the amount of contraction. A band of paper or a revolving drum-surface moves at a uniform speed at right angles to the direction of motion of the writing lever. When the muscle recovers from the stimulus, it relaxes into its original form, and the writing point traces the recovery as it moves now to the left, regaining its first position. A curve is thus described, the rising portion of which is due to contraction, and the falling portion to relaxation or recovery. The ordinate of the curve represents the intensity of response, and the abscissa the time (*Figure 1*).

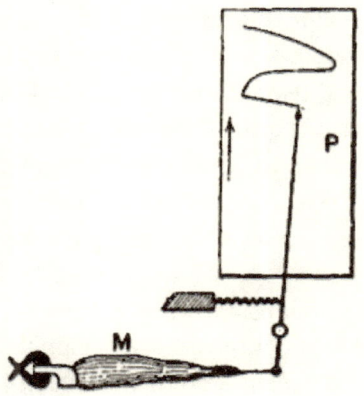

Fig. 1 – **Mechanical lever recorder**

The muscle M with the attached bone is securely held at one end, the other end being connected with the writing lever. Under the action of stimulus the contracting muscle pulls the lever and moves the tracing point to the right over the travelling recording surface P. When the muscle recovers from contraction, the tracing point returns to its original position. See on P the record of muscle curve.

## Characteristics of the response curve: (1) period, (2) amplitude, (3) form

Just as a wave of sound is characterised by its (1) period, (2) amplitude, and (3) form, so may these response curves be distinguished from each other. As regards the period, there is an enormous variation, corresponding to the functional activity of the muscle. For instance, in tortoise it may be as high as a second, whereas in the wing muscles of many insects it is as small as 1/300 part of a second.

> "It is probable that a continuous graduated scale might, as suggested by Hermann, be drawn up in the animal kingdom, from the excessively rapid contraction of insects to those of tortoises and hibernating dormice." [1]

Differences in form and amplitude of curve are well illustrated by various muscles of the tortoise. The curve for the muscle of the neck, used for rapid withdrawal of the head on approach of danger, is quite different from that of the pectoral muscle of the same animal, used for its sluggish movements.

Again, progressive changes in the same muscle are well seen in the modifications of form which consecutive muscle-curves gradually undergo. In a dying muscle, for example, the amplitude of succeeding curves is continuously diminished, and the curves themselves are elongated. Numerous illustrations will be seen later, of the effect, in changing the form of the curve, of the increased excitation or depression produced by various agencies.

Thus these response records give us a means of studying the effect of stimulus, and the modification of response, under varying external conditions, advantage being taken of the mechanical contraction produced in the tissue by the stimulus. But there are other kinds of tissue where the excitation produced by stimulus is not exhibited in a visible form. In order to study these we have to use an altogether independent method; the method of *electric response*.

## Footnotes

1. Biedermann, *Electro-physiology*, p. 59.

# Chapter 2
# ELECTRIC RESPONSE

*Conditions for obtaining electric response – Method of injury – Current of injury – Injured end, cuproid: uninjured, zincoid – Current of response in nerve from more excited to less excited – Difficulties of present nomenclature – Electric recorder – Two types of response, positive and negative – Universal applicability of electric mode of response – Electric response a measure of physiological activity – Electric response in plants.*

Unlike muscle, a length of nerve, when mechanically or electrically excited, does not undergo any visible change. That it is thrown into an excitatory state, and that it conducts the excitatory disturbance, is shown however by the contraction produced in an attached piece of muscle, which serves as an indicator.

But the excitatory effect produced in the nerve by stimulus can also be detected by an electrical method. If an isolated piece of nerve be taken and two contacts be made on its surface by means of non-polarisable electrodes at A and B, connection being made with a galvanometer, no current will be observed, as both A and B are in the same physico-chemical condition. The two points, that is to say, are iso-electric.

If now the nerve be excited by stimulus, similar disturbances will be evoked at both A and B. If, further, these disturbances, reaching A and B almost simultaneously, cause any electrical change, then, similar changes taking place at both points, and there being thus no relative difference between the two, the galvanometer will still indicate no current. This null-effect is due to the balancing action of B as against A. (See *Figure 2a*)

## Conditions for obtaining electric response

If then we wish to detect the response by means of the galvanometer, one means of doing so will lie in the abolition of this balance, which may be accomplished by making one of the two points, say B, more or less permanently irresponsive. In that case, stimulus will cause greater electrical disturbance at the more responsive point, say A, and this will be shown by the galvanometer as a current of response. To make B less responsive we may injure it by means of a cross-sectional cut, a burn, or the action of strong chemical reagents.

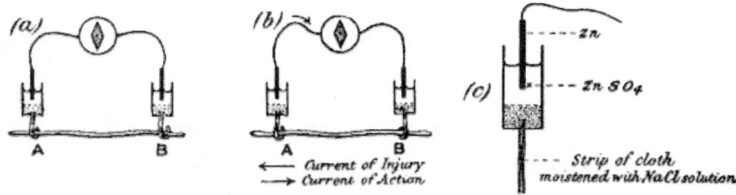

Fig. 2 **Electric method of detecting nerve response**

(a) Iso-electric contacts; no current in the galvanometer.

(b) The end **B** injured; current of injury from **B** to **A**: stimulation gives rise to an action current from **A** to **B**.

(c) Non-polarisable electrode.

## Current of injury

We shall revert to the subject of electric response; meanwhile it is necessary to say a few words regarding the electric disturbance caused by the injury itself. Since the physico-chemical conditions of the uninjured A and the injured B are now no longer the same, it follows that their electric conditions have also become different. They are no longer iso-electric. There is thus a more or less permanent or resting difference of electric potential between them. A current – the current of injury – is found to flow *in the nerve*, from the injured to the uninjured, and in the galvanometer, through the electrolytic contacts from the uninjured to the injured. As long as there is no further disturbance this current of injury remains approximately constant, and is therefore sometimes known as 'the current of rest' (*Figure 2b*).

A piece of living tissue, unequally injured at the two ends, is thus seen to act like a voltaic element, comparable to a copper and zinc couple. As some confusion has arisen, on the question of whether the injured end is like the zinc or copper in such a combination, it will perhaps be well to enter upon this subject in detail.

If we take two rods, of zinc and copper respectively, in metallic contact, and further, if the points A and B are connected by a strip of cloth *s* moistened with salt solution, it will be seen that we have a complete voltaic element. A current will now flow from B to A in the metal (*Figure 3a*) and from A to B through the electrolyte *s*. Or instead of connecting A and B by a single strip of cloth *s*, we may connect them by two strips *s s'* leading to non-polarisable electrodes E E'. The current will then be found just the same as before, i.e. from B to A in the metallic part, and from A through *s s'* to B, the wire W being interposed, as it were, in the electrolytic part of the circuit. If now a galvanometer be interposed at O, the current will flow from B to A through the galvanometer, i.e. from right to left. But if we interpose the galvanometer in the electrolytic part of the circuit, that is to say, at W, the same current will appear to flow in the opposite direction. In *Figure 3c*, the galvanometer is so interposed, and in this case it is to be noticed that when the current in the galvanometer flows from left to right, the metal connected to the left is zinc.

Compare *Figure 3d*, where A B is a piece of nerve of which the B end is injured. The current in the galvanometer through the non-polarisable electrode is from left to right. The uninjured end is therefore comparable to the zinc in a voltaic cell (i.e it is zincoid), the injured being copper-like or cuproid.[2]

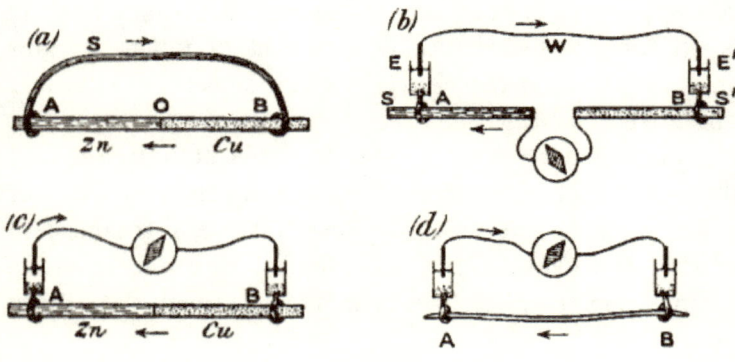

**Fig. 3 Diagram showing the correspondence between injured (b) and uninjured (a) contacts in nerve, and Cu and Zn in a voltaic element**

Comparison of *(c)* and *(d)* will show that the injured end of B in *(d)* corresponds with the Cu in *(c)*.

If the electrical condition of, say, zinc in the voltaic couple *Figure 3c*) undergoes any change (and I shall show later that this can be caused by molecular disturbance), then the existing difference of potential between A and B will also undergo variation. If for example the electrical condition of A approaches that of B, the potential difference will undergo a diminution, and the current hitherto flowing in the circuit will, as a consequence, display a diminution, or *negative* variation.

## Action current

We have seen that a current of injury – sometimes known as 'current of rest' – flows in a nerve from the injured to the uninjured, and that the injured B is then less excitable than the uninjured A. If now the nerve be excited, there being a greater effect produced at A, the existing difference of potential may thus be reduced, with a consequent diminution of the current of injury. During stimulation, therefore, a nerve exhibits a negative variation. We may express this in a different way by saying that a 'current of action' was produced in response to stimulus, and acted in an opposite direction to the current of injury (*Figure 2b*). The action current in the nerve *is from the relatively more excited* to the *relatively less excited*.

## Difficulties of present nomenclature

We shall deal later with a method by which a responsive current of action is obtained without any antecedent current of injury. 'Negative variation' has then no meaning. Or, again, a current of injury may sometimes undergo a change of direction (see note 4, p.42). In view of these considerations it is necessary to have at our disposal other forms of expression by which the direction of the current of response can still be designated.

Keeping in touch with the old phraseology, we might then call a current 'negative' that which flows from the *more* excited to the *less* excited. Or, bearing in mind the fact that an uninjured contact acts as the zinc in a voltaic couple, we might call it 'zincoid,' and the injured contact 'cuproid.'

Stimulation of the uninjured end, approximating it to the condition of the injured, might then be said to induce a 'cuproid change'.

The electric change produced in a normal nerve by stimulation may therefore be expressed by saying that there has been a negative variation, or that there was a current of action from the more excited to the less excited, or that stimulation has produced a cuproid change.

The excitation, or molecular disturbance, produced by a stimulus has thus a concomitant electrical expression. As the excitatory state disappears with the return of the excitable tissue to its original condition, the current of action will gradually disappear.[3] The movement of the galvanometer needle during excitation of the tissue thus indicates a molecular upset by the stimulus; and the gradual creeping back of the galvanometer deflection exhibits a molecular recovery.

This transitory electrical variation constitutes the 'response,' and its intensity varies according to that of the stimulus.

## Electric recorder

We have thus a method of obtaining curves of response electrically. After all, it is not essentially very different from the mechanical method. In this case we use a magnetic lever (*Figure 4a*), the needle of the galvanometer, which is deflected by the electromagnetic pull of the current, generated under the action of stimulus, just as the mechanical lever was deflected by the mechanical pull of the muscle contracting under stimulus.

The accompanying diagram (*Figure 4b*) shows how, under the action of stimulus, the current of rest undergoes a transitory diminution, and how on the cessation of stimulus there is gradual recovery of the tissue, as exhibited in the return of the galvanometer needle to its original position.

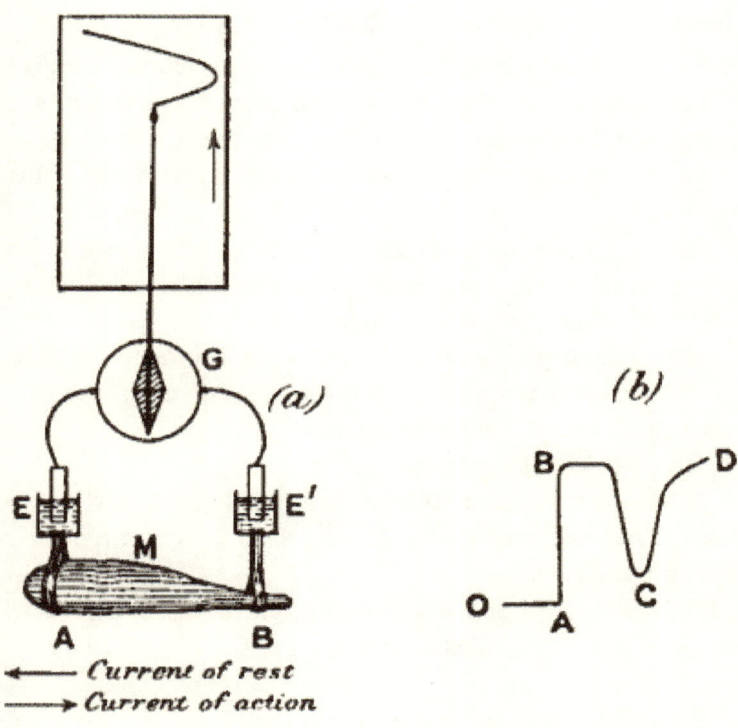

Fig. 4 **Electric recorder**

(a) M muscle; A uninjured, B injured ends. E E' non-polarising electrodes connecting A and B with galvanometer G. Stimulus produces negative variation of current of rest. Index connected with galvanometer needle records curve on travelling paper (in practice, moving galvanometer spot of light traces curve on photographic plate). Rising part of curve shows effect of stimulus; descending part, recovery.

(b) O is the zero position of the galvanometer; injury produces a deflection A B; stimulus diminishes this deflection to C; C D is the recovery.

## Two types of response – positive and negative

It may here be added that though stimulus in general produces a diminution of current of rest, or a negative variation (e.g. muscles and nerves), yet, in certain cases, there is an increase, or positive variation. This is seen in the response of the retina to light.

Again, a tissue which normally gives a negative variation may undergo molecular changes, after which it gives a positive variation. Thus Dr. Waller finds that whereas fresh nerve always gives negative variation, stale nerve sometimes gives positive; and that retina, which when fresh gives positive, when stale, exhibits negative variation.

The following is a summary of the two types of response:

1) *Negative variation* – Action current from more excited to less excited – cuproid change in the excited – e.g. fresh muscle and nerve, stale retina.

2) *Positive variation* – Action current from less excited to more excited – zincoid change in the excited – e.g. stale nerve, fresh retina.[4]

From this it will be seen that it is the fact of the electrical response of living substances to stimulus that is of essential importance, the sign *plus* or *minus* being a minor consideration.

## Universal applicability of the electrical mode of response

This mode of obtaining electrical response is applicable to all living tissues, and in cases like that of muscle, where mechanical response is also available, it is found that the electrical and mechanical records are practically identical.

The two response curves seen in the accompanying diagram *(Figure 5)*, and taken from the same muscle by the two methods simultaneously, clearly exhibit this. Thus we see that electrical response can not only take the place of the mechanical record, but has the further advantage of being applicable in cases where the latter cannot be used.

## Electrical response: A measure of physiological activity

These electrical changes are regarded as physiological, or characteristic of living tissue, for any conditions which enhance physiological activity also, *pari passu*, increase their intensity. Again, when the tissue is killed by poison, electrical response disappears, the tissue passing into an irresponsive condition. Anæsthetics, like chloroform, gradually diminish, and finally altogether abolish, electrical response.

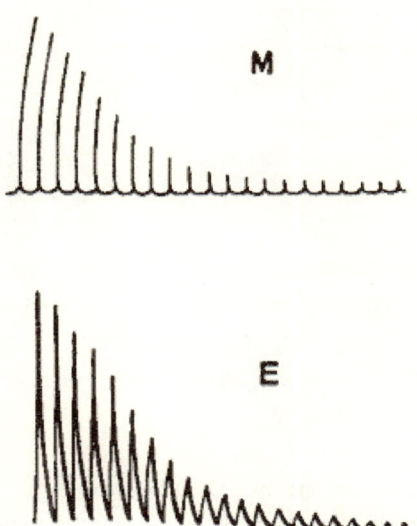

Fig. 5 **Simultaneous record of the mechanical (M) and (E) electrical responses of the muscle of frog** (Waller)

From these observed facts – that living tissue gives response while a tissue that has been killed does not – it is concluded that the phenomenon of response is peculiar to living organisms.[5] The response phenomena that we have been studying are therefore considered as due to some unknown, super-physical 'vital' force and are thus relegated to a region beyond physical inquiry.

It may, however, be that this limitation is not justified,

and surely, at least until we have explored the whole range of physical action, it cannot be asserted definitely that a particular class of phenomena is by its very nature outside that category.

## Electric response in plants

But before we proceed to the inquiry as to whether these responses are or are not due to some physical property of matter, and are to be met with even in inorganic substances, it will perhaps be advisable to see whether they are not paralleled by phenomena in the transitional world of plants. We shall thus pass from a study of response in highly complex animal tissues to those given under simpler vital conditions.

Electric response has been found by Munck, Burdon-Sanderson, and others to occur in sensitive plants. But it would be interesting to know whether these responses were confined to plants which exhibit such remarkable mechanical movements, and whether they could not also be obtained from ordinary plants where visible movements are completely absent. In this connection, Kunkel observed electrical changes in association with the injury or flexion of stems of ordinary plants.[6]

My own attempt, however, was directed, not towards the obtaining of a mere qualitative response, but rather to the determination of whether throughout the whole range of response phenomena a parallel between animal and vegetable could be detected. That is to say, I desired to know, with regard to plants, what was the relation between intensity of stimulus and the corresponding response; what were the effects of superposition of stimuli; whether fatigue was present, and in what manner it influenced response; what were the effects of extremes of temperature on the response; and, lastly, if chemical reagents could exercise any influence in the modification of plant response, as stimulating, anæsthetic, and poisonous drugs have been found to do with nerve and muscle.

If it could be proved that the electric response served as a faithful index of the physiological activity of plants, it would then be possible successfully to attack many problems in plant physiology, the solution of which at present offers many experimental difficulties.

With animal tissues, experiments have to be carried on under many great and unavoidable difficulties. The isolated tissue, for example, is subject to unknown changes inseparable from the rapid approach of death. Plants, however, offer a great advantage in this respect, for they maintain their vitality unimpaired during a very great length of time.

In animal tissues, again, the vital conditions themselves are highly complex. Those essential factors which modify response can, therefore, be better determined under the simpler conditions which obtain in vegetable life.

In the following chapters it will be shown that the response phenomena are exhibited not only by plants but by inorganic substances as well, and that the responses are modified by various conditions in exactly the same manner as those of animal tissues. In order to show how striking are these similarities, I shall for comparison place side by side the responses of animal tissues and those I have obtained with plants and inorganic substances. For the electric response in animal tissues, I shall take the latest and most complete examples from the records made by Dr. Waller.

But before we can obtain satisfactory and conclusive results regarding plant response, many experimental difficulties will have to be surmounted. I shall now describe how this has been accomplished.[7]

## Footnotes

2. In some physiological text-books much wrong inference has been made, based on the supposition that the injured end is zinc-like.

3. "The exciting cause is able to produce a particular molecular rearrangement in the nerve; this constitutes the state of excitation and is accompanied by local electrical changes as an ascertained physical concomitant.

   The excitatory state evoked by stimulus manifests itself in nerve fibres by E.M. changes, and as far as our present knowledge goes by these only. The conception of such an excitable living tissue as nerve implies that of a molecular state which is in stable equilibrium. This equilibrium can be readily upset by an external agency, the stimulus, but the term "stable" expresses the fact that a change in any direction must be succeeded by one of opposite character, this being the return of the living structure to its previous state. Thus the electrical manifestation of the excitatory state is one whose duration depends upon the time during which the external agent is able to upset and retain in a new poise the living equilibrium, and if this is extremely brief, then the recoil of the tissue causes such manifestation to be itself of very short duration." – *Text-book of Physiology*, ed. Schäfer, ii. 453.

4. I shall here mention briefly one complication that might arise from regarding the current of injury as the current of reference, and designating the response current either positive or negative in relation to it. If this current of injury remained always invariable in direction – that is to say, from the injured to the uninjured – there would be no source of uncertainty. But it is often found, for example in the retina, that the current of injury undergoes a reversal, or is reversed from the beginning. That is to say, the direction is now from the uninjured to the injured, instead of the opposite. Confusion is thus very apt to arise. No such misunderstanding can however occur if we call the current of response towards the more excited *positive*, and towards the less excited *negative*.

5. "The Electrical Sign of Life ... An isolated muscle gives sign of life by contracting when stimulated ... An ordinary nerve, normally connected with its terminal organs, gives sign of life by means of muscle, which by direct or reflex path is set in motion when the nerve trunk is stimulated. But such nerve separated from its natural termini, isolated from the rest of the organism, gives no sign of life when excited, either in the shape of chemical or of thermic changes, and it is only by means of an electrical change that we can ascertain whether or no it is alive ... The most general and most delicate sign of life is then the electrical response." – Waller, in *Brain*, pp. 3 and 4. Spring 1900.

6. Kunkel thought the electric disturbance to be due to movement of water through the tissue. It will be shown that this explanation is inadequate.

7. My assistant Mr. J. Bull has rendered me very efficient help in these experiments.

## Chapter 3
# ELECTRIC RESPONSE IN PLANTS – METHOD OF NEGATIVE VARIATION

*Negative variation – Response recorder – Photographic recorder – Compensator – Means of graduating intensity of stimulus – Spring tapper and torsional vibrator – Intensity of stimulus dependent on amplitude of vibration – Effectiveness of stimulus dependent on rapidity also.*

I shall first proceed to show that an electric response is evoked in plants under stimulation.[8]

In experiments for the exhibition of electric response it is preferable to use a non-electrical form of stimulus, for there is then a certainty that the observed response is entirely due to reaction from stimulus, and not, as might be the case with electric stimulus, to mere escape of stimulating current through the tissue. For this reason, the mechanical form of stimulation is the most suitable.

I find that all parts of the living plant give electric response to a greater or less extent. Some, however, give stronger response than others. In favourable cases, we may have an E.M. variation as high as .1 volt. It must however be remembered that the response, being a function of physiological activity of the plant, is liable to undergo changes at different seasons of the year. Each plant has its particular season of maximum responsiveness. The leaf stalk of horse chestnut, for example, exhibits fairly strong response in spring and summer, but on the approach of autumn it undergoes diminution. I give here a list of specimens which will be found to exhibit fairly good response:

**Root:** Carrot (*Daucus Carota*), radish (*Raphanus sativus*).

**Stem:** Geranium (*Pelargonium*), vine (*Vitis vinifera*).

**Leaf stalk:** Horse-chestnut (*Aesculus Hippocastanum*), turnip (*Brassica Napus*), cauliflower (*Brassica oleracea*), celery (*Apium graveolens*), Eucharis lily (*Eucharis amazonica*).

**Flower stalk:** Arum lily (*Richardia africana*).

**Fruit:** Eggplant (*Solanum Melongena*).

**Negative variation:** Taking the leaf stalk of turnip we kill an area on its surface, say B, by the application of a few drops of strong potash, the area at A being left uninjured. A current is now observed to flow, in the stalk, from the injured B to the uninjured A, as was found to be the case in the animal tissue. The potential difference depends on the condition of the plant, and the season in which it may have been gathered. In the experiment here described (*Figure 6a*) its value was 0.13 volt.

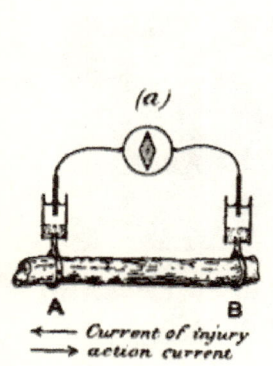
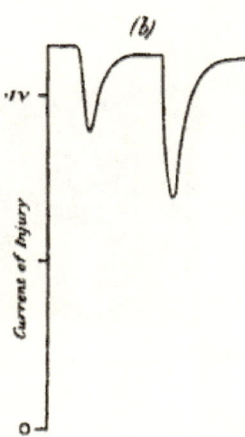

Fig. 6

*(a)* Experiment for exhibiting electric response in plants by method of negative variation. *(b)* Responses in leaf stalk of turnip to stimuli of two successive taps, the second being stronger.

> A and B contacts are about 2 cm. apart, B being injured. Plant is stimulated by a tap between A and B. Stimulus acts on both A and B, but owing to injury of B, effect at A is stronger and a negative variation due to differential action occurs.

A sharp tap was now given to the stalk, and a sudden diminution, or negative variation, of current occurred, the resting potential difference being decreased by .026 volt. A second and stronger tap produced a second response, causing a greater diminution of P.D. by .047 volt (*Figure 6b*). The accompanying figure is a photographic record of another set of response curves (*Figure 7*). The first three responses are for a given intensity of stimulus, and the next six in response to stimulus nearly twice as strong. It will be noticed that fatigue is exhibited in these responses. Other experiments will be described in the next chapter which show conclusively that the response was not due to any accidental circumstance but was a direct result of stimulation. But I shall first discuss the experimental arrangements and method of obtaining these graphic records.

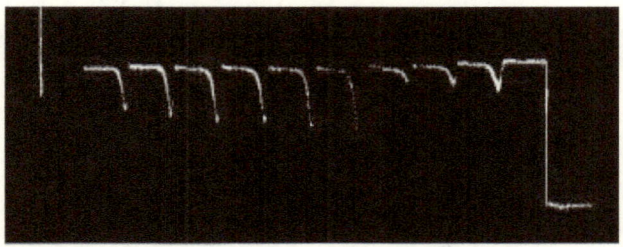

Fig. 7 **Record of responses in plant (leaf stalk of cauliflower) by method of negative variation**

The first three records are for stimulus intensity 1; the next six are for intensity twice as strong; the successive responses exhibit fatigue. The vertical line to the left represents 0·1 volt. The record is to be read from right to left.

## Response recorder

The galvanometer used is a sensitive dead-beat D'Arsonval. The period of complete swing of the coil under experimental conditions is about 11 seconds. A current of $10^{-9}$ ampere produces a deflection of 1 mm. at a distance of 1 metre.

For a quick and accurate method of obtaining the records, I devised the following form of response recorder.

The curves are obtained directly, by tracing the excursion of the galvanometer spot of light on a revolving drum *(Figure 8)*. The drum, on which is wrapped the paper for receiving the record, is driven by clockwork. Different speeds of revolution can be given to it by adjustment of the clock governor, or by changing the size of the driving wheel. The galvanometer spot is thrown down on the drum by the inclined mirror M. The galvanometer deflection takes place at right angles to the motion of the paper. A stylographic pen attached to a carrier rests on the writing surface. The carrier slides over a rod parallel to the drum. As has been said before, the galvanometer deflection takes place parallel to the drum, and as long as the plant rests unstimulated, the pen, remaining coincident with the stationary galvanometer spot on the revolving paper, describes a straight line.

If, on stimulation, we trace the resulting excursion of the spot of light, by moving the carrier which holds the pen, the rising portion of the response curve will be obtained. The galvanometer spot will then return more or less gradually to its original position, and that part of the curve which is traced during the process constitutes the recovery.

The ordinate in these curves represents the E.M. variation, and the abscissa the time.

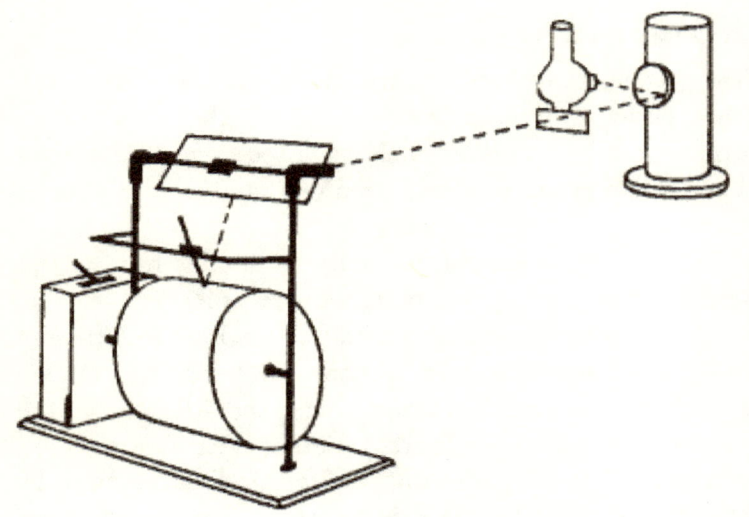

Fig. 8 **Response recorder**

We can calibrate the value of the deflection by applying a known E.M.F. to the circuit from a compensator, and noting the deflection which results. The speed of the clock is previously adjusted so that the recording surface moves exactly through, say, one inch a minute. Of course this speed can be increased to suit the particular experiment, and in some it is as high as six inches a minute. In this simple manner very accurate records may be made. It has the additional advantage that one is able at once to see whether the specimen is suitable for the purpose of investigation. A large number of records might be taken by this means in a comparatively short time.

## Photographic recorder

Or the records may be made photographically. A clockwork arrangement moves a photographic plate at a known uniform rate, and a curve is traced on the plate by the moving spot of light. All the records that will be given are accurate reproductions of those obtained by one of these two methods.

*Photographic records are reproduced in white against a black background.*

## Compensator

As the responses are on *variation* of current of injury, and as the current of injury may be strong, and throw the spot of light beyond the recording surface, a potentiometer balancing arrangement may be used (*Figure 9*), by which the P.D. due to injury is exactly compensated; E.M. variations produced by stimulus are then taken in the usual manner. This compensating arrangement is also helpful, as has been said before, for calibrating the E.M. value of the deflection.

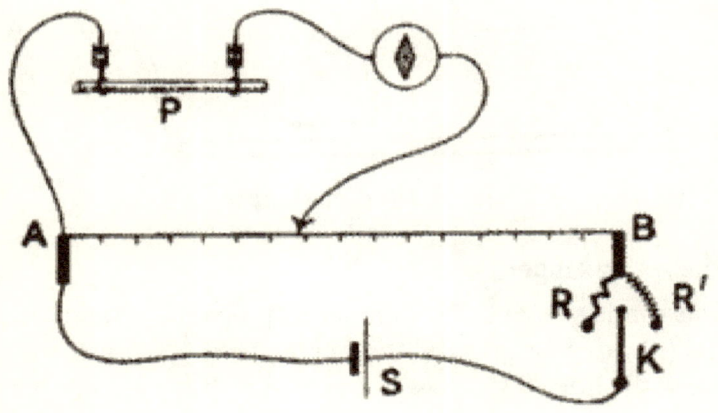

Fig. 9 **The compensator**

A B is a stretched wire with added resistances R and R'. S is a storage cell. When the key K is turned to the right one scale division = .001 volt, when turned to the left one scale division = .01 volt. P is the plant.

## Means of graduating the intensity of stimulus

One of the necessities in connection with quantitative measurements is to be certain that the intensity of successive stimuli is (1) constant, or (2) capable of gradual increase by known amounts. No two taps given by the hand can be made exactly alike. I have therefore devised the two following methods of stimulation, which have been found to act satisfactorily.

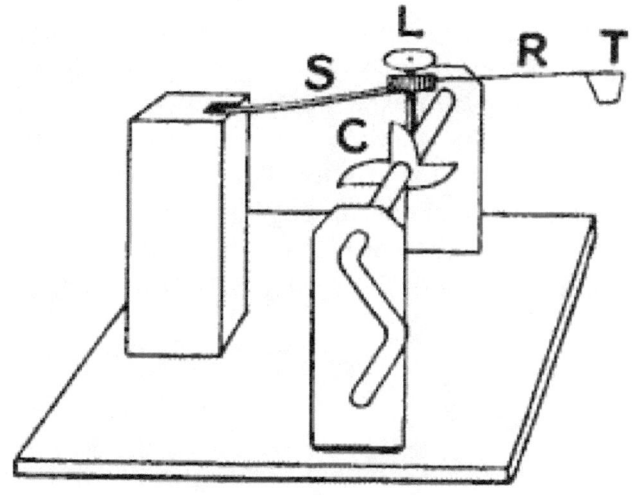

Fig. 10 **The spring tapper**

## The spring tapper

This consists (*Figure 10*) of the spring proper (S), the attached rod (R) carrying at its end the tapping-head (T). A projecting rod – the lifter (L) – passes through S R. It is provided with a screw thread, by means of which its length, projecting downwards, is regulated. This fact, as we shall see, is made to determine the height of the stroke. (C) is a cogwheel. As one of the spokes of the cogwheel is rotated past (L), the spring is lifted and released, and (T) delivers a sharp tap. The height of the lift, and therefore the intensity of the stroke, is measured by means of a graduated scale. We can increase the intensity of the stroke through a wide range (1) by increasing the projecting length of the lifter, and (2) by shortening the length of spring by a sliding catch.

We may give isolated single taps or superpose a series in

rapid succession according as to whether the wheel is rotated slow or fast. The only disadvantage of the tapping method of stimulation is that in long experiments the point struck is liable to be injured. The vibrational mode of stimulation to be presently described labours under no such disadvantage.

## The electric tapper

Instead of the simple mechanical tapper, an electromagnetic tapper may be used.

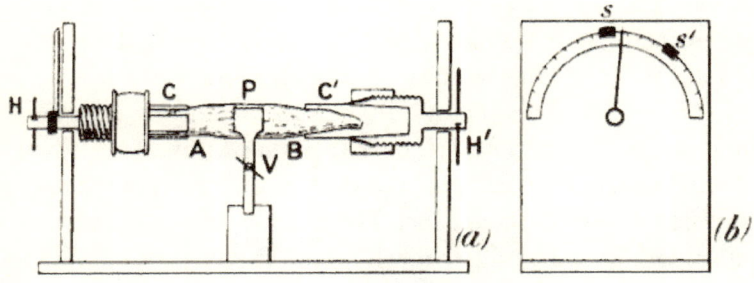

Fig. 11 **The torsional vibrator**

Plant **P** is securely held by a vice **V**. The two ends are clamped by holders **C C'**. By means of handles **H H'**, torsional vibration may be imparted to either the end **A** or end **B** of the plant. The end view (*b*) shows how the amplitude of vibration is predetermined by means of movable stops **S S'**.

## Vibrational stimulus

I find that torsional vibration affords another very effective method of stimulation (*Figure 11*). The plant stalk may be fixed in a vice (V), the free ends being held in tubes (C C'), provided with three clamping jaws. A rapid torsional vibration[9] may now be imparted to the stalk by means of the handle (H). The amplitude of vibration, which determines the intensity of stimulus, can be accurately measured by the graduated circle. The amplitude of vibration may be predetermined by means of the sliding stops (S S').

## Intensity of stimulus dependent on amplitude of vibration

I shall now describe an experiment which shows that torsional vibration is as effective as stimulation by taps, and that its stimulating intensity increases, length of stalk being constant, with amplitude of vibration.

It is of course obvious that if the length of the specimen be doubled, the vibration, in order to produce the same effect, must be through twice the angle.

I took a leaf stalk of turnip and fixed it in the torsional vibrator. I then recorded responses to two successive taps, the intensity of one being nearly double that of the other. Having done this, I applied to the same stalk two successive torsional vibrations of 45° and 67° respectively. These successive responses to taps and torsional vibrations are given in *Figure 12*, and from them it will be seen that these two modes of stimulation may be used indifferently, with equal effect. The vibrational method has the advantage over tapping, that, while with the latter the stimulus is somewhat localised, with vibration the tissue subjected to stimulus is uniformly stimulated throughout its length.

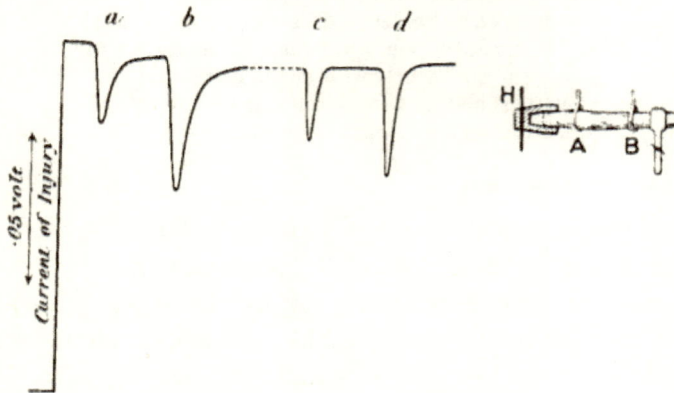

Fig. 12 **Response in plant to mechanical tap or vibration**

The end **B** is injured. A tap was given between **A** and **B** and this gave the response curve *a*. A stronger tap gave the response *b*. By means of the handle **H**, a torsional vibration of 45° was now imparted, this gave the response *c*. Vibration through 67° gave *d*.

## Effectiveness of stimulus dependent on rapidity also

In order that successive stimuli may be equally effective another point has to be borne in mind. In all cases of stimulation of living tissue it is found that the effectiveness of a stimulus to arouse response depends on the rapidity of the onset of the disturbance. It is thus found that the stimulus of the 'break' induction shock, on a muscle for example, is more effective, by reason of its greater rapidity, than the 'make' shock. So also with the torsional vibrations of plants, I find response depending on the quickness with which the vibration is effected. I give below records of successive stimuli, given by vibrations through the same amplitude, but delivered with increasing rapidity (*Figure 13*).

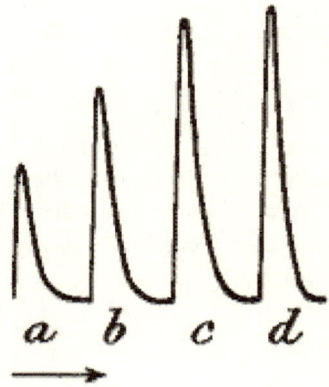

**Fig. 13 Influence of suddenness on the efficiency of stimulus**

The curves *a*, *b*, *c*, *d*, are responses to vibrations of the same amplitude, 30°. In *a* the vibration was very slow; in *b* it was less slow; it was rapid in *c*, and very rapid in *d*.

Thus if we wish to maintain the effective intensity of stimulus constant we must meet two conditions: (1) The amplitude of vibration must be kept the same. This is done by means of the graduated circle. (2) The vibration period must be kept the same. With a little practice, this requirement is easily fulfilled.

The uniformity of stimulation which is thus attained solves the great difficulty of obtaining reliable quantitative values, by whose means alone can rigorous demonstration of the phenomena we are studying become possible.

## Footnotes

8. A preliminary account of electric response in plants was given at the end of my paper on *Electric Response of Inorganic Substances* read before the Royal Society on June 6, 1901; also at the Friday evening discourse, Royal Institution, May 10, 1901. A more complete account is given in my paper on *Electric Response in Ordinary Plants under Mechanical Stimulus* read before the Linnean Society March 20, 1902.

    I thank the Royal Society and the Linnean Society for permission to reproduce some of my diagrams published in their *Proceedings*. – J. C. B.

9. By this is meant a rapid to-and-fro or complete vibration. In order that successive responses should be uniform it is essential that there should be no resultant twist, i.e. the plant at the end of vibration should be in exactly the same condition as at the beginning.

# Chapter 4
# ELECTRIC RESPONSE IN PLANTS – BLOCK METHOD

*Method of block – Advantages of block method*
*– Plant response a physiological phenomenon*
*– Abolition of response by anæsthetics and poisons*
*– Abolition of response when plant is killed by hot water.*

I shall now proceed to describe another and independent method which I devised for obtaining plant response. It has the advantage of offering us a complementary means of verifying the results found by the method of negative variation. As it is also, in itself, for reasons which will be shown later, a more perfect mode of inquiry, it enables us to investigate problems which would otherwise have been difficult to attempt.

When electrolytic contacts are made on the uninjured surfaces of the stalk at A and B, the two points, being practically similar in every way, are iso-electric, and little or no current will flow in the galvanometer. If now the whole stalk be uniformly stimulated, and if both ends A and B be equally excited at the same moment, it is clear that there will still be no responsive current, owing to balancing action at the two ends. This difficulty as regards the obtaining of response was overcome in the method of negative variation, where the excitability of one end was depressed by chemical reagents or injury, or abolished by excessive temperature. On stimulating the stalk there was produced a greater excitation at A than at B, and a current of action was then observed to flow in the stalk from the more excited A to the less excited B (*Figure 6*).

But we can cause this differential action to become evident by another means. For example, if we produce a block, by clamping at C between A and B (*Figure 14a*), so that the disturbance made at A by tapping or vibration is prevented from reaching B, we shall then have A thrown into a relatively greater excitatory condition than B. It will now be found that a current of action flows in the stalk from A to B, that is to say, from the excited to the less excited. When the B end is stimulated, there will be a reverse current (*Figure 14b*).

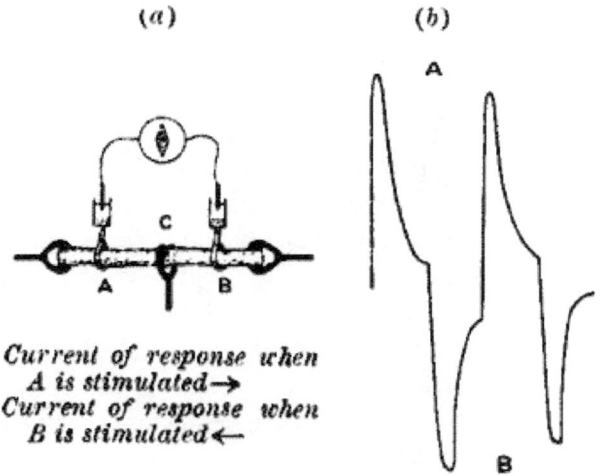

Fig. 14 **The method of block**

*(a)* The plant is clamped at C, between A and B.

*(b)* Responses obtained by alternately stimulating the two ends. Stimulation of A produces upward response; of B gives downward response.

We have in this method a great advantage over that of negative variation, for we can always verify any set of results by making corroborative reversal experiments.

By the method of injury again, one end is made initially abnormal, i.e. different from the condition which it maintains when intact. Further, inevitable changes will proceed unequally at the injured and uninjured ends, and the conditions of the experiment may thus undergo unknown variations. But by the block method which has just been described, there is no injury, the plant is normal throughout, and any physiological change (which in plants will be exceedingly small during the time of the experiment) will affect it as a whole.

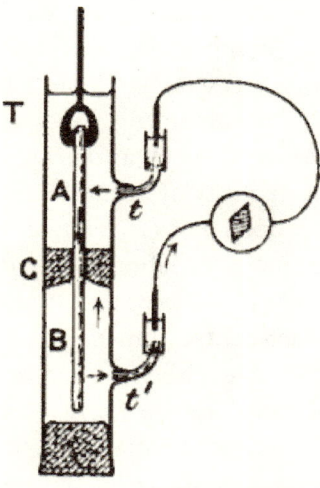

**Fig. 15 Response in plant (from the stimulated A to unstimulated B) completely immersed under water**

The leaf stalk is clamped securely in the middle with the cork C, inside the tube T, which is filled with water, the plant being completely immersed. Moistened threads in connection with the two non-polarisable electrodes are led to the side tubes t t'. One end of the stalk is held in ebonite forceps and vibrated. A current of response is found to flow in the stalk from the excited A to the unexcited B, and outside, through the liquid, from B to A. A portion of this current, flowing through the side tubes t t', produces deflection in the galvanometer.

## Plant response a physiological or vital response

I now proceed to a demonstration of the fact that whatever be the mechanism by which they are brought about, these plant responses are physiological in their character. As the investigations described in the next few chapters will show, they furnish an accurate index of physiological activity. For it will be found that, other things being equal, whatever tends to exalt or depress the vitality of the plant tends also to increase or diminish its electric response. These E.M. effects are well marked, and attain considerable value, rising sometimes, as has been said before, to as much as 0.1 volt or more. They are proportional to the intensity of stimulus.

It need hardly be added that special precautions are taken to avoid shifting of contacts. Variation of contact, however, could not in any case account for repeated transient responses to repeated stimuli, when contact is made on iso-electric surfaces. Nor could it in any way explain the reversible nature of these responses, when A and B are stimulated alternately. These responses are obtained in the plants even when completely immersed in water, as in the experimental arrangement (*Figure 15*). It will be seen that in this case, where there could be no possibility of shifting of contact, or variation of surface, there is still the usual current of response.

I shall describe here a few crucial experiments only, in proof of the physiological character of electric response. The test applied by physiologists, in order to discriminate as to the physiological nature of response, consists in finding out whether the response is diminished or abolished by the action of anæsthetics, poisons, and excessively high temperature, which are known to depress or destroy vitality.

I shall therefore apply these same tests to plant responses.

## Effect of anæsthetics and poisons

Ordinary anæsthetics, like chloroform, and poisons, like mercuric chloride, are known to produce a profound depression or abolish all signs of response in the living tissue. For the purpose of experiment, I took two groups of stalks, with leaves attached, exactly similar to each other in every respect. In order that the leaf stalks might absorb chloroform I dipped their cut ends in chloroform-water, a certain amount of which they absorbed, the process being helped by the transpiration from the leaves. The second group of stalks was placed simply in water, in order to serve as the experiment controls. The narcotic action of chloroform, finally culminating in death, soon became visually evident. The leaves began to droop, and a peculiar death-discolouration began to spread from the mid rib along the venation of the leaves.

Another peculiarity was also observed. The aphides feeding on the leaves died even before the appearance of the discoloured patches, whereas on the leaves of the stalks placed in water these little creatures maintained their accustomed activity, nor did any discolouration occur. In order to study the effect of poison, another set was placed in water containing a small quantity of mercuric chloride. The leaves here underwent the same change of appearance, and the aphides met with the same untimely fate, as in the case of those subjected to the action of chloroform. There was hardly any visible change in the appearance of the stalks themselves, which were to all outer appearances as living as ever, indications of death being apparent only on the leaf surfaces.

I give below the results of several sets of experiments, from which it would appear that whereas there was strong normal response in the group of stalks kept in water, there was practically a total abolition of all response in those anæsthetised or poisoned.

## Experiments on the effect of anæsthetics and poisons

A batch of ten leaf stalks of a plane tree was placed with the cut ends in water, and leaves in air; an equal number was immersed in chloroform-water; a third batch was placed in a 5% solution of mercuric chloride.

Similarly a batch of three horse-chestnut leaf stalks was put in water, another batch in chloroform-water, and a third batch in mercuric chloride solution.

| 1) Leaf stalk of Plane tree<br>The stimulus applied was a single vibration of 90° | | | | | |
|---|---|---|---|---|---|
| a) after 24 hours in water<br>(All leaves standing up and fresh – aphides alive) | | B) after 24 hours in chloroform water<br>(Leaves began to droop in 1 hour and bent over in 3 hours – aphides dead) | | C) after 24 hours in mercuric chloride<br>(Leaves began to droop in 4 hours. Deep discolouration along the veins. Aphides dead) | |
| | electric response | | electric response | | electric response |
| (1) | 21 dn | (1) | 1 dn | (1) | 0 dn |
| (2) | 31 dn | (2) | 1 dn | (2) | 0.25 dn |
| (3) | 26 dn | (3) | 2 dn | (3) | 0.25 dn |
| (4) | 15 dn | (4) | 0 dn | (4) | 0 dn |
| (5) | 17 dn | (5) | 1 dn | (5) | 0.25 dn |
| (6) | 23 dn | (6) | 1.5 dn | (6) | 0.25 dn |
| (7) | 30 dn | (7) | 2 dn | (7) | 0 dn |
| (8) | 27 dn | (8) | 1 dn | (8) | 0.25 dn |
| (9) | 29 dn | (9) | 1 dn | (9) | 0.25 dn |
| (10) | 17 dn | (10) | 0.5 dn | (10) | 0.5 dn |
| Mean response 23.6 | | Mean 1 | | Mean 0.15 | |
| 2) Leaf stalk of Horse Chestnut | | | | | |
| (1) | 15 | (1) | 0.5 | (1) | 0 |
| (2) | 17 | (2) | 0.5 | (2) | 0 |
| (3) | 10 | (3) | 0 | (3) | 0 |
| Mean 14 | | Mean 0.3 | | Mean 0 | |

These results conclusively prove the physiological nature of the response.

I shall in a succeeding chapter give a continuous series of response curves showing how, owing to progressive death from the action of poison, the responses undergo steady diminution till they are completely abolished.

## Effect of high temperature

It is well known that plants are killed when subjected to high temperatures. I took a stalk, and, using the block method, with torsional vibration as the stimulus, obtained strong responses at both ends A and B. I then immersed the same stalk for a short time in hot water at about 65°C, and again stimulated it as before. But at neither A nor B could any response now be evoked. As all the external conditions were the same in the first and second parts of this experiment, the only difference being that in one the stalk was alive and in the other killed, we have here further and conclusive proof of the physiological character of electric response in plants.

The same facts may be demonstrated in a still more striking manner by first obtaining two similar but opposite responses in a fresh stalk, at A and B, and then killing one half, say B, by immersing only that half of the stalk in hot water. The stalk is replaced in the apparatus, and it is now found that whereas the A half gives strong response, the end B gives none.

In the experiments on negative variation, it was tacitly assumed that the variation is due to a differential action, stimulus producing a greater excitation at the uninjured than at the injured end. The block method enables us to test the correctness of this assumption. The B end of the stalk is injured or killed by a few drops of strong potash, the other end being uninjured. There is a clamp between A and B. The end A is stimulated and a strong response is obtained. The end B is now stimulated, and there is little or no response. The block is now removed and the plant stimulated throughout its length. Though the stimulus now acts on both ends, yet, owing to the irresponsive condition of B, there is a resultant response, which from its direction is found to be due to the responsive action of A. This would not have been the case if the end B had been uninjured. We have thus experimentally verified the assumption that in the same tissue an uninjured portion will be thrown into a greater excitatory state than an injured, by the action of the same stimulus.

## Chapter 5

# PLANT RESPONSE – ON THE EFFECTS OF SINGLE STIMULUS AND OF SUPERPOSED STIMULI

*Effect of single stimulus – Superposition of stimuli – Additive effect – Staircase effect – Fatigue – No fatigue when sufficient interval between stimuli – Apparent fatigue when stimulation frequency is increased – Fatigue under continuous stimulation.*

## Effect of single stimulus

In a muscle a single stimulus gives rise to a single twitch which may be recorded either mechanically or electrically. If there is no fatigue, the successive responses to uniform stimuli are exactly similar. Muscle when strongly stimulated often exhibits fatigue, and successive responses therefore become feebler and feebler. In nerves, however, there is practically no fatigue and successive records are alike. Similarly, in plants, we shall find some exhibiting marked fatigue and others very little.

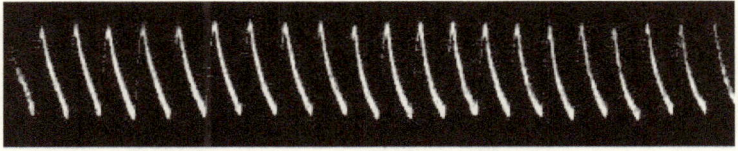

Fig. 16 **Uniform responses (radish)**

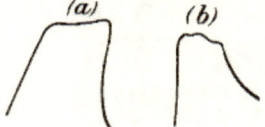

Fig. 17 **Fusion of effect of rapidly succeeding stimuli**

*(a)* in muscle; *(b)* in carrot.

## Superposition of stimuli.

If instead of a single stimulus a succession of stimuli is superposed, it happens that a second shock is received before recovery from the first has taken place. Individual effects will then become more or less fused.

When the frequency is sufficiently increased, the intermittent effects are fused, and we find an almost unbroken curve.

When for example the muscle attains its maximum contraction (corresponding to the frequency and strength of stimuli) it is thrown into a state of complete *tetanus*, in which it appears to be held rigid. If the rapidity is not sufficient for this, we have the jagged curve of incomplete tetanus. If there is not much fatigue, the upper part of the tetanic curve is approximately horizontal, but in cases where fatigue sets in quickly, the fact is shown by the rapid decline of the curve.

With regard to all these points we find strict parallels in plant response. In cases where there is no fatigue, the successive responses are identical (*Figure 16*). With superposition of stimuli we have fusion of effects, analogous to the tetanus of muscle (*Figure 17*). And lastly, the influence of fatigue in plants is to produce a modification of response curve exactly similar to that of muscle (see below). One effect of superposition of stimuli may be mentioned here.

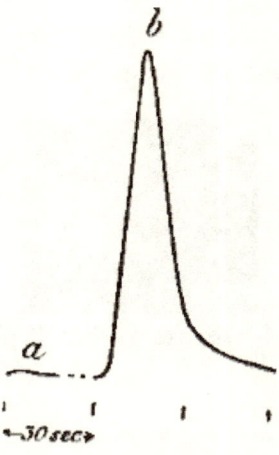

Fig. 18 **Additive Effect**

A single stimulus of 3° vibration produced little or no effect *(a)*, but the same stimulus when rapidly superposed thirty times produced the large effect *(b)*. (Leaf stalk of turnip.)

## Additive effect

It is found in animal responses that there is a minimum intensity of stimulus, below which no response can be evoked. But even a sub-minimal stimulus will, though singly ineffective, become effective when repeated. In plants, too, we obtain a similar effect, i.e. the summation of single ineffective stimuli produces effective response (*Figure 18*).

## Staircase effect

Animal tissues sometimes exhibit what is known as the 'staircase effect,' that is to say, the heights of successive responses are gradually increased, though the stimuli are maintained constant. This is exhibited typically by cardiac muscle, though it is not unknown even in nerve.

The cause is obscure, but it seems to depend on the condition of the tissue. It appears as if the molecular sluggishness of tissue were in these cases only gradually removed under stimulation, and the increased effects were due to increased molecular mobility. Whatever the explanation, I have sometimes observed the same staircase effect in plants (*Figure 19*).

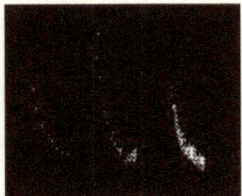

Fig. 19 **'Staircase effect' in plant**

## Fatigue

It is assumed that in living substances like muscle, fatigue is caused by the break down or dissimilation of tissue by stimulus. And till this waste is repaired by the process of building-up or assimilation, the functional activity of the tissue will remain below par. There may also be an accumulation of the products of dissimilation – 'the fatigue stuffs' – and these latter may act as poisons or chemical depressants.

In an animal it is supposed that the nutritive blood supply performs the two-fold task of bringing material for assimilation and removing the fatigue products, thus causing the disappearance of fatigue.

This explanation, however, is shown to be insufficient by the fact that an excised bloodless muscle recovers from fatigue after a short period of rest. It is obvious that here the fatigue has been removed by means other than that of renewed assimilation and removal of fatigue products by the circulating blood. It may therefore be instructive to study certain phases of fatigue exhibited under simpler conditions in vegetable tissue, where the constructive processes are in abeyance, and there is no active circulation for the removal of fatigue products.

It has been said before that the E.M. variation caused by stimulus is the concomitant of a disturbance of the molecules of the responsive tissues from their normal equilibrium, and that the curve of recovery exhibits the restoration of the tissue to equilibrium.

## No fatigue when sufficient interval between successive stimuli

We may thus gather from a study of the response curve some indication of the molecular distortion experienced by the excited tissue. Let us first take the case of an experiment whose record is given in *Figure 20a*. It will be seen from that curve that one minute after the application of stimulus there is a complete recovery of the tissue; the molecular condition is exactly the same at the end of recovery as in the beginning of stimulation. The second and succeeding response curves therefore are exactly similar to the first, *provided a sufficient interval has been allowed in each case for complete recovery*. There is, in such a case, no diminution in intensity of response, that is to say, there is no fatigue.

We have an exactly parallel case in muscles.

> "In muscle with normal circulation and nutrition there is always an interval between each pair of stimuli, in which the height of twitch does not diminish even after protracted excitation, and no fatigue appears." [10]

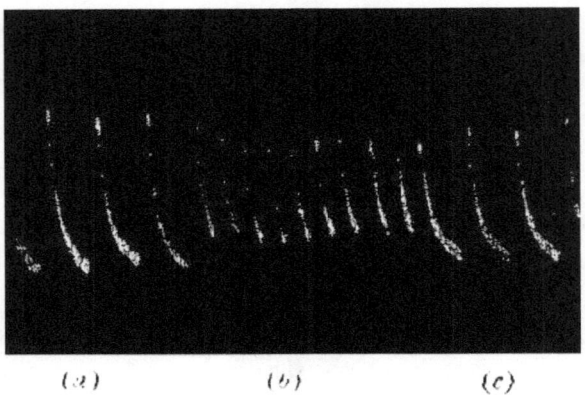

(a)  (b)  (c)

Fig. 20 **Record showing diminution of response when sufficient time is not allowed for full recovery**

In *(a)* stimuli were applied at intervals of one minute; in *(b)* the intervals were reduced to half a minute; this caused a diminution of response. In *(c)* the original rhythm is restored, and the response is found to be enhanced. (Radish.)

## Apparent fatigue when stimulation frequency increased.

If the rhythm of stimulation frequency is now changed, and made quicker, certain remarkable modifications will appear in the response curves. In *Figure 20*, the first part shows the responses at one minute intervals, by which time the individual recovery was complete.

The rhythm was now changed to intervals of half a minute, instead of one, while the stimuli were maintained at the same intensity as before. It will be noticed (*Figure 20b*) that these responses appear much feebler than the first set, in spite of the equality of stimulus. An inspection of the figure may perhaps throw some light on the subject. It will be seen that when greater frequency of stimulation was introduced, the tissue had not yet had time to effect complete recovery from previous strain. The molecular swing towards equilibrium had not yet abated, when the new stimulus, with its opposing impulse, was received. There is thus a diminution of height in the resultant response. The original rhythm of one minute was now restored, and the succeeding curves (*Figure 20c*) at once show increased response. An analogous instance may be cited in the case of muscle response, where 'the height of twitch diminishes more rapidly in proportion as the excitation interval is shorter.'[11]

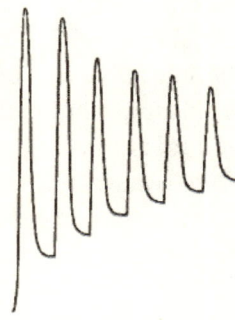

Fig. 21 **Fatigue in celery**

Vibration of 30° at intervals of half a minute.

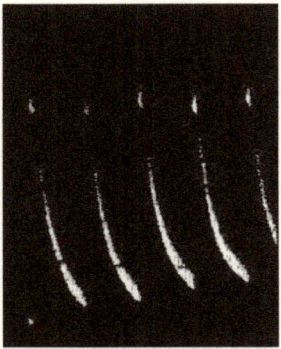

Fig. 22 **Fatigue in leaf stalk of cauliflower**

Stimulus: 30° vibration at intervals of one minute.

From what has just been said it would appear that one of the causes of diminution of response, or fatigue, is the residual strain. This is clearly seen in *Figure 21*, in a record which I obtained with celery stalk. It will be noticed there that, owing to the imperfect molecular recovery during the time allowed, the succeeding heights of the responses have undergone a continuous diminution. *Figure 22* gives a photographic record of fatigue in the leaf stalk of cauliflower.

It is evident that residual strain, other things being equal, will be greater if the stimuli have been excessive. This is well seen in *Figure 23*, where the set of first three curves A is for stimulus intensity of 45° vibration, and the second set B, with an augmented response, for stimulus intensity of 90° vibration. On reverting in C to stimulus intensity of 45°, the responses are seen to have undergone a great diminution as compared with the first set A. Here is seen marked fatigue, the result of overstrain from excessive stimulation.

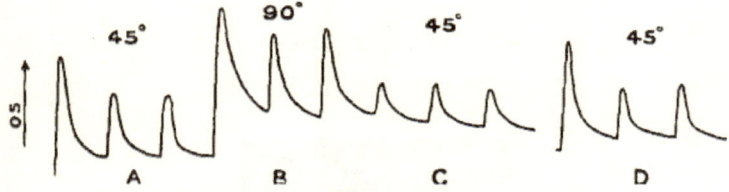

Fig. 23 **Effect of overstrain in producing fatigue**

Successive stimuli applied at intervals of one minute. The intensity of stimulus in C is the same as that of A, but response is feebler owing to previous over-stimulation. Fatigue is to a great extent removed after fifteen minutes' rest, and the responses in D are stronger than those in C. The vertical line between arrows represents .05 volt. (Turnip leaf stalk.)

If this fatigue is really due to residual strain effect, then, as strain disappears with time, we may expect the responses to regain their former height after a period of rest. In order to verify this, therefore, I renewed the stimulation (at intensity 45°) after fifteen minutes. It will at once be seen from record D how far the fatigue had been removed.

One peculiarity that will be noticed in these curves is that, owing to the presence of comparatively little residual strain, the first response of each set is relatively large. The succeeding responses are approximately equal where the residual strains are similar. The first response of A shows this because it had had long previous rest. The first of B shows it because we are there seeing for the first time increased stimulation. The first of C does *not* show it, because there is now a strong residual strain. D again shows it because the strain has been removed by fifteen minutes' rest.

### Fatigue under continuous stimulation

The effect of fatigue is exhibited in marked degree when a tissue is subjected to continuous stimulation. In cases where there is marked fatigue, as for instance in certain muscles, the top of the tetanic curve undergoes rapid decline. A similar effect is obtained also with plants (*Figure 24*).

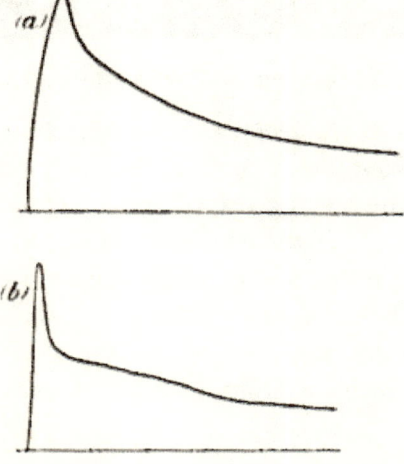

Fig. 24 **Rapid fatigue under continuous stimulation in (a) muscle; (b) in leaf stalk of celery**

The effect of rest in producing molecular recovery, and hence in the removal of fatigue, is well illustrated in the following set of photographic records (*Figure 25*). The first shows the curve obtained with a fresh plant. The effect is seen to be very large. Two minutes were allowed for recovery, and then stimulation was repeated during another two minutes. The response in this case is seen to be decidedly smaller. A third case is somewhat similar to the second. A period of rest of five minutes was now allowed, and the curve obtained subsequently, owing to partial removal of residual strain, is found to exhibit greater response.

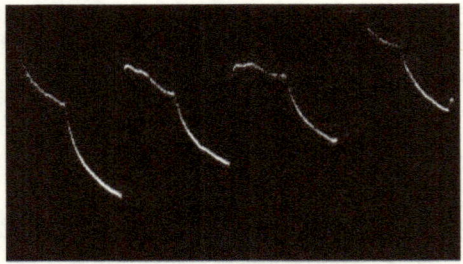

Fig. 25 **Effect of continuous vibration (through 50°) in carrot**

In the first three records, two minutes' stimulation is followed by two minutes' recovery. The last record was taken after the specimen had a rest of five minutes. The response, owing to removal of fatigue by rest, is stronger.

The results thus arrived at, under the simple conditions of vegetable life, free as they are from all possible complications and uncertainties, may perhaps throw some light on the obscure phenomena of fatigue in animal tissues.

## Footnotes

10. Biedermann, *Electro-physiology*, p. 86.
11. Biedermann, *loc. cit.*

# Chapter 6
# PLANT RESPONSE – ON DIPHASIC VARIATION

*Diphasic variation – Positive after-effect and positive response – Radial E.M. variation.*

When a plant is stimulated at any point, a molecular disturbance – the excitatory wave – is propagated outwards from the point of its initiation.

## Diphasic variation

This wave of molecular disturbance is attended by a wave of electrical disturbance. (Usually speaking, the electrical relation between disturbed and less disturbed is that of copper to zinc.)

It takes some time for a disturbance to travel from one point to another, and its intensity may undergo a diminution as it recedes further from its point of origin. Suppose a disturbance originated at C; if two points are taken near each other, as A and B, the disturbance will reach them almost at the same time, and with the same intensity. The electric disturbance will be the same in both. The effect produced at A and B will balance each other and there will be no resultant current.

By killing or otherwise reducing the sensibility of B as is done in the method of injury, there is no response at B, and we obtain the unbalanced response, due to disturbance at A; the same effect is obtained by putting a clamp between A and B, so that the disturbance may not reach B. But we may get response even without injury or block.

If we have the contacts at A and B, and if we give a tap *nearer*

A than B (*Figure 26a*), then we have:

(1) the disturbance reaching A earlier than B.

(2) The disturbance reaching A is much stronger than at B. The disturbance at B may be so comparatively feeble as to be negligible.

It will thus be seen that we might obtain responses even without injury or block, in cases where the disturbance is enfeebled in reaching a distant point. In such a case on giving a tap near A a responsive current would be produced in one direction, and in the opposite direction when the tap is given near B (*Figure 26b*). Theoretically, then, we might find a neutral point between A and B, so that, on originating the disturbance there, the waves of disturbance would reach A and B at the same instant and with the same intensity. If, further, the rate of recovery be the same for both points, then the electric disturbances produced at A and B will continue to balance each other, and the galvanometer will show no current. On taking a cylindrical root of radish I have sometimes succeeded in finding a neutral point, which, being disturbed, did not give rise to any resultant current. But disturbing a point to the right or to the left gave rise to opposite currents.

It is, however, difficult to obtain an absolutely cylindrical specimen, as it always tapers in one direction. The conductivity towards the tip of the root is not exactly the same as that in the ascending direction. It is therefore difficult to fix an absolutely neutral point, but a point may be found which approaches this very nearly, and on stimulating the stalk near this, a very interesting diphasic variation has been observed. In a specimen of cauliflower stalk:

(1) stimulus was applied very much nearer A than B (the feeble disturbance reaching B was negligible). The resulting response was upward and the recovery took place in about sixty seconds.

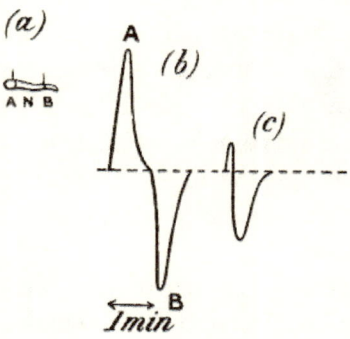

Fig. 26 **Diphasic variation**

(2) Stimulus was next applied near B. The resulting response was now downward (*Figure 26b*).

(3) The stimulus was now applied near the approximately neutral point N. In this case, owing to a slight difference in the rates of propagation in the two directions, a very interesting diphasic variation was produced (*Figure 26c*). From the record it will be seen that the disturbance arrived earlier at A than at B. This produced an upward response. But during the subsidence of the disturbance at A, the wave reached B. The effect of this was to produce a current in the opposite direction. This apparently hastened the recovery of A (from 60 seconds to 12 seconds). The excitation of A now disappeared, and the second phase of response, that due to excitation of B, was fully displayed.

## Positive after-effect

If we regard the response due to excitation of A as negative, the later effect on B would appear as a subsequent positive variation.

In the response of nerve, for example, where contacts are made at two surfaces, injured and uninjured, there is sometimes observed, first a negative variation, and then a positive after-effect. This may sometimes at least be due to the proximal uninjured contact first giving the usual negative variation, and the more distant contact of injury giving rise, later, to the opposite, that is to say, apparently positive, response. There is always a chance of an after-effect due to this cause, unless (1) the injured end be completely killed and rendered quite irresponsive, or (2) there be an effective block between A and B, so that the disturbance due to stimulus can only act on one, and not on the other.

I have found cases where, even when there was a perfect block, a positive after-effect occurred. It would thus appear that if molecular distortion from stimulus gives rise to a negative variation, then during the process of molecular recovery there may be over-shooting of the equilibrium position, which may be exhibited as a positive variation.

## Positive variation

The responses given by muscle or nerve are, normally speaking, negative. But that of retina is positive. The sign of response, however, is apt to be reversed if there is any molecular modification of the tissue from changes of external circumstances. Thus it is often found that nerve in a stale condition gives positive, instead of the normal negative variation, and stale retina often gives negative, instead of the usual positive.

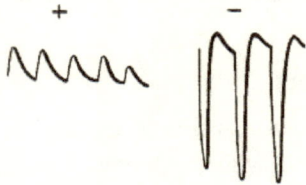

Fig. 27 **Abnormal positive responses in stale leaf stalk of turnip converted into normal negative under strong stimulation**[12]

The relative intensities of stimuli in the two cases are in the ratio of 1:7.

Curiously enough, I have on many occasions found exactly parallel instances in the response of plants. Plants when fresh, as stated, give negative responses as a rule. But when somewhat faded they sometimes give rise to positive response. Again, just as in the modified nerve the abnormal positive response gives place to the normal negative under strong and long-continued stimulation, so also in the modified plant the abnormal positive response passes into negative (*Figure 27*) under strong stimulation. I was able in some cases to trace this process of gradual reversal, by continuously increasing the intensity of stimulus. It was then found that as the stimulus was increased, the positive at a certain point underwent a reversal into the normal negative response (*Figure 28*).

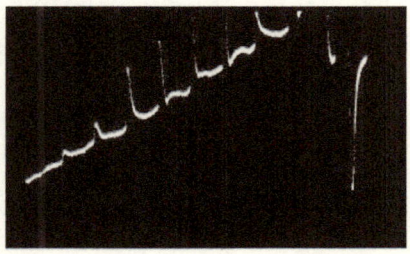

**Fig. 28 Abnormal positive passing into normal negative in a stale specimen of leaf stalk of cauliflower**

Stimulus was gradually increased from 1 to 10, by means of spring tapper. When the stimulus intensity was 10, the response became reversed into normal negative. (Parts of 8 and 9 are out of the plate.)

The plant thus gives a reversed response under abnormal conditions of staleness. I have sometimes found similar reversal of response when the plant is subjected to the abnormal conditions of excessively high or low temperature.

## Radial E.M. variation

We have seen that a current of response flows in the plant from the relatively more to the relatively less excited. A theoretically important experiment is the following:

A thick stem of plant stalk was taken and a hole bored so as to make one contact with the interior of the tissue, the other being on the surface. After a while the current of injury was found to disappear. On exciting the stem by taps or torsional vibration, a responsive current was observed which flowed inwards from the more disturbed outer surface to the shielded core inside (*Figure 29*).

Hence it is seen that when a wave of disturbance is propagated along the plant, there is a concomitant wave of radial E.M. variation. The swaying of a tree by the wind would thus appear to give rise to a radial E.M.F.

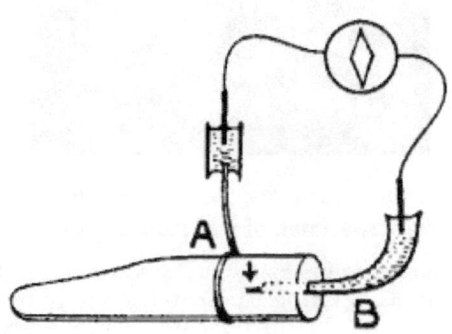

Fig. 29 **Radial E.M. variation**

## Footnotes

12. For general purposes it is immaterial whether the responses are recorded up or down. For convenience of inspection they are in general recorded up. But in cases where it is necessary to discriminate the sign of response, positive response will be recorded up, and negative down.

# Chapter 7
# PLANT RESPONSE – ON THE RELATION BETWEEN STIMULUS AND RESPONSE

*Increased response with increasing stimulus*
*– Apparent diminution of response with excessively strong stimulus.*


As already said, in the living tissue, molecular disturbance induced by stimulus is accompanied by an electric disturbance, which gradually disappears with the return of the disturbed molecules to their position of equilibrium. The greater the molecular distortion produced by the stimulus, the greater the electric variation produced. The electric response is thus an outward expression of a molecular disturbance produced by an external agency, i.e the stimulus.

## Curve of relation between stimulus and response

In the curve showing the relation between stimulus and response in nerve and muscle, it is found that the molecular effect as exhibited either by contraction or E.M. variation in muscle, or simply by E.M. variation in nerve, is at first slight. In the second part, there is a rapidly increasing effect with increased stimulus. Finally, a tendency shows itself to approach a limit of response. Thus we find the curve at first slightly convex, then straight and ascending, and lastly, concave to the abscissa (*Figure 30*).

In muscle the limit of response is reached much sooner than in nerve. As will be seen, the range of variation of stimulus in these curves is not very great. When the stimulus is carried beyond moderate limits, the response, owing to fatigue or other causes, may sometimes undergo an actual diminution.

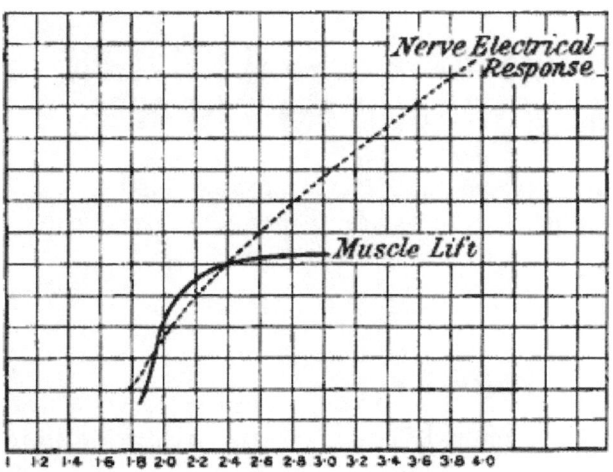

Fig. 30 **Curves showing the relation between the intensity of stimulus and response**

Abscissæ indicate increasing intensity of stimulus. Ordinates indicate magnitude of response. (Waller.)

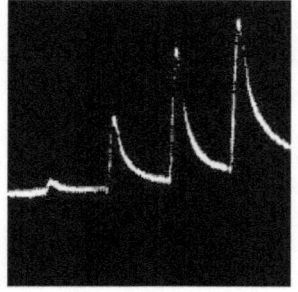

Fig. 31

Taps of increasing strength 1:2:3:4 producing increased response in leaf stalk of turnip.

I have obtained very interesting results, with reference to the relation between stimulus and response, when experimenting with plants. These results are suggestive of various types of response met with in animal tissues.

1) In order to obtain the simplest type of effects, not complicated by secondary phenomena, one has to choose specimens which exhibit little fatigue. Having procured these, I undertook two series of experiments. In the first (A) the stimulus was applied by means of the spring tapper, and in the second (B) by torsional vibration.

(A) The first stimulus was given by a fall of the lever through $h$, the second through $2h$, and so on. The response curves clearly show increasing effect with increased stimulus (*Figure 31*).

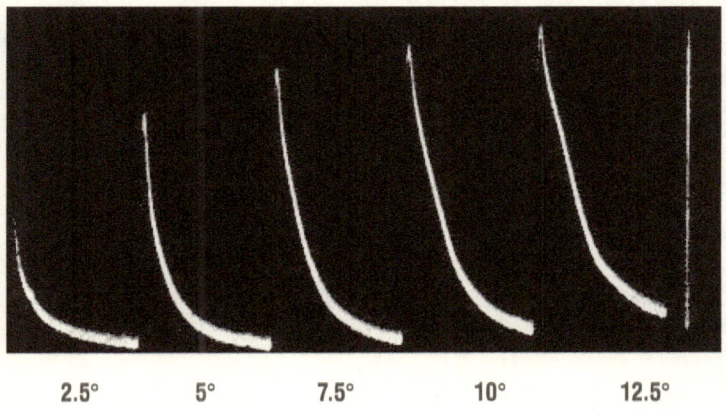

2.5°    5°    7.5°    10°    12.5°

### Fig. 32 Increased response with increasing vibrational stimuli (cauliflower stalk)

Stimuli applied at intervals of three minutes.
Vertical line = 0·1 volt.

(B) 1) The vibrational stimulus was increased from 2.5° to 5° to 7.5° to 10° to 12.5° in amplitude. It will be observed how the intensity of response tends to approach a limit (*Figure 32*).

## Table showing the increased E.M. variation produced by increasing stimulus

| angle of vibration | e.m.f |
|---|---|
| 2.5° | .044 volt |
| 5° | .075 volt |
| 7.5° | .090 volt |
| 10° | .100 volt |
| 12.5° | .106 volt |

2) The next figure shows how little variation is produced with a low value of stimulus, but with increasing stimulus the response undergoes a rapid increase, after which it tends to approach a limit (*Figure 33a*).

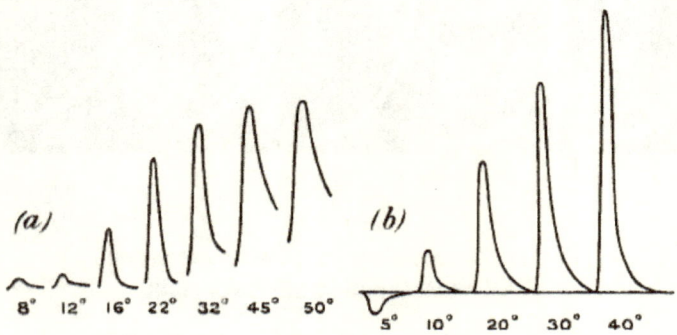

Fig. 33 **Responses to increasing stimuli produced by increasing angle of vibration**

(*a*) Record with a specimen of fresh radish. Stimuli applied at intervals of two minutes. The record is taken for one minute.

(*b*) Record for stale radish. There is a reversed response for the feeble stimulus of 5° vibration.

3) As an extreme instance of the case just cited, I have often come across a curious phenomenon. During the gradual increase of the stimulus from a low value there would be

apparently no response. But when a critical value was reached a maximum response would suddenly occur, and would not be exceeded when the stimulus was further increased. Here we have a parallel to what is known in animal physiology as the 'all or none' principle. With the cardiac muscle, for example, there is a certain minimal intensity which is effective in producing response, but further increase of stimulus produces no increase in response.

4) From an inspection of the records of responses which are given, it will be seen that the slope of a curve which shows the relation of stimulus to response will at first be slight, the curve will then ascend rapidly, and at high values of stimulus tend to become horizontal. The curve as a whole becomes first slightly convex to the abscissa, then straight and ascending, and lastly concave. A far more pronounced convexity in the first part is shown in some cases, especially when the specimen is stale.

This is due to the fact that under these circumstances response is apt to begin with an actual reversal of sign, the plant under feebler than a certain critical intensity of stimulus giving positive, instead of the normal negative, response (*Figure 33b*).

## Diminution of response with excessively strong stimulus

It is found that in animal tissues there is sometimes an actual diminution of response with excessive increase of stimulus. Thus Waller finds, in working with retina, that as the intensity of light stimulus is gradually increased, the response at first increases, and then sometimes undergoes a diminution. This phenomenon is unfortunately complicated by fatigue, itself regarded as obscure. It is therefore difficult to say whether the diminution of response is due to fatigue or to some reversing action of an excessively strong stimulus.

From *Figure 33b*, above, it is seen that there was an actual reversal of response in the lower portion of the curve. It is therefore not improbable that there may be more than one point of reversal.

In physical phenomena we are, however, acquainted with numerous instances of reversals. For example, a common effect of magnetisation is to produce an elongation of an iron rod. But Bidwell finds that as the magnetising force is pushed to an extreme, at a certain point elongation ceases and is succeeded, with further increase of magnetising force, by an actual contraction. Again a photographic plate, when exposed continuously to light, gives at first a negative image. Still longer exposure produces a positive. Then again we have a negative. There is thus produced a series of recurrent reversals. In photographic prints of flashes of lightning, two kinds of images are observed, one, the positive – when the lightning discharge is moderately intense – and the other, negative, the so-called 'dark lightning' – due to the reversal action of an intensely strong discharge.

In studying the changes of conductivity produced in metallic particles by the stimulus of Hertzian radiation, I have often noticed that whereas feeble radiation produces one effect, strong radiation produces the opposite. Again, under the continuous action of electric radiation, I have frequently found recurrent reversals.[13]

## Diminution of response under strong stimulus traced to fatigue

But there are instances in plant response where the diminution effect can be definitely traced to fatigue. The records of these cases are extremely suggestive as to the manner in which the diminution is brought about.

The accompanying figures (*Figure 34*) give records of responses to increasing stimulus. They were made with specimens of cauliflower stalks, one of which (*a*) showed little fatigue, while in the other (*b*) fatigue was present. It will be seen that the curves obtained by joining the apices of the successive single responses are very similar.

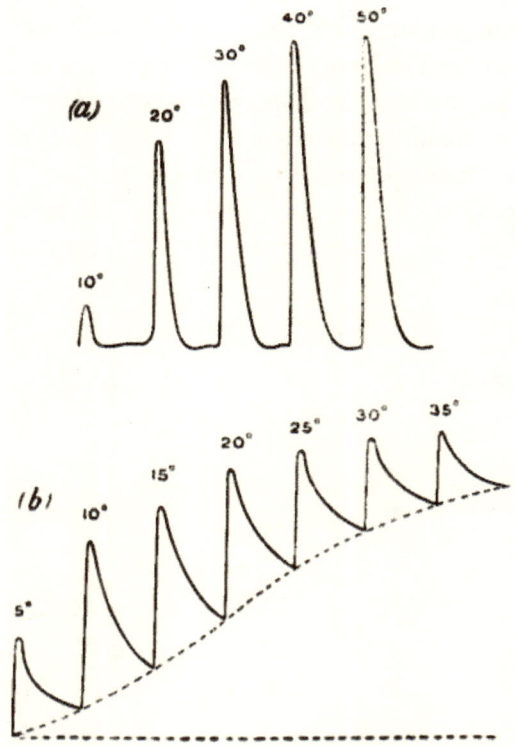

Fig. 34 **Responses to increasing stimulus obtained with two specimens of stalk of cauliflower**

In *(a)* fatigue is absent, in *(b)* it is present.

In one case there is no fatigue, the recovery from each stimulus being complete. Every response in the series therefore starts from a position of perfect equilibrium, and the height of the single responses increases with increasing stimulation. But in the second case, the strain is not completely removed after any single stimulation of the series. That recovery is partial is seen by the gradual shifting of the base line upwards. In the former case the base line is horizontal and represents a condition of complete equilibrium. Now, however, the base line, or line of modified equilibrium, is tilted upwards.

Thus even in this case if we measure the heights of successive responses from the line of absolute equilibrium, they will be found to increase with increasing stimulus. Ordinarily, however, we make no allowance for the shifting of the base line, measuring response rather from the place of its previous recovery, or from the point of modified equilibrium. Judged in this way, the responses undergo an apparent diminution.

## Footnotes

13. See *On Electric Touch*, Proc. Roy. Soc. Aug. 1900.

# Chapter 8
# PLANT RESPONSE – ON THE INFLUENCE OF TEMPERATURE

*Effect of very low temperature – Influence of high temperature – Determination of death point – Increased response as after-effect of temperature variation – Death of plant and abolition of response by the action of steam.*

For every plant there is a range of temperature most favourable to its vital activity. Above this optimum, the vital activity diminishes, till a maximum is reached, when it ceases altogether, and if this point is maintained for a long time the plant is apt to be killed. Similarly, the vital activity is diminished if the temperature is lowered below the optimum, and again, at a minimum point it ceases, while below this minimum the plant may be killed. We may regard these maximum and minimum temperatures as the 'death points'. Some plants can resist these extremes better than others.

Length of exposure, it should however be remembered, is also a determining factor in the question as to whether or not the plant can survive unfavourable conditions of temperature.

Thus we have hardy plants, and plants that are affected by excessive variations of temperature. Within each species, there may be, again, a certain amount of difference between individual plants.

These facts being known, I was anxious to determine whether the undoubted changes induced by temperature in the vital activity of plants would affect electrical response.

## Effect of very low temperature

As regards the influence of very low temperature, I had the opportunity of studying the question with the sudden appearance of frost.

In the previous week, when the temperature was about 10°C, I had obtained strong electric response in radishes whose value varied from 0.05 to 0.1 volt. But two or three days later, with the effect of the frost, I found electric response to have practically disappeared. A few radishes were, however, found somewhat resistant, but the electric response had, even in these cases, fallen from the average value of 0.075 V. under normal temperature to 0.003 V. after the frost. That is to say, the average sensitiveness had been reduced to about 1/25$^{th}$.

On warming the frost-bitten radish to 20°C there was an appreciable revival, as shown by increase in response. In specimens where the effect of frost had been very great, i.e. in those which showed little or no electric response, warming did not restore responsiveness.

From this it would appear that frost killed some, which could not be subsequently revived, whereas others were only reduced to a condition of torpidity, from which there was revival on warming.

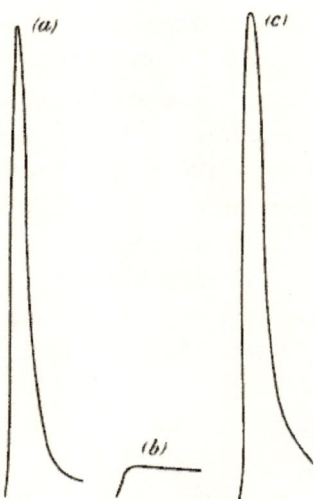

**Fig. 35 Diminution of response in eucharis by lowering of temperature**

*(a)* Normal response at 17°C.

*(b)* The response almost disappears when the plant is subjected to −2°C for fifteen minutes.

*(c)* Revival of response on warming to 20°C.

I now tried the effect of artificial lowering of temperature on various plants. A plant which is very easily affected by cold is a certain species of Eucharis lily. I first obtained responses with the leaf stalk of this lily at the ordinary temperature of the room (17°C). I then placed it for fifteen minutes in a cooling chamber, temperature −2°C, for only ten minutes, after which, on trying to obtain response, it was found to have practically disappeared. I now warmed the plant by immersing it for a while in water at 20°C, and this produced a revival of the response (*Figure 35*). If the plant is subjected to low temperature for too long a time, there is then no subsequent revival.

I obtained a similar marked diminution of response with the flower stalk of Arum lily, on lowering the temperature to zero.

My next attempt was to compare the sensibility of different plants to the effect of lowered temperatures. For this purpose I chose three specimens: (1) Eucharis lily; (2) Ivy; and (3) Holly. I took their normal response at 17°C, and found that, generally speaking, they attained a fairly constant value after the third or fourth response. After taking these records of normal response, I placed the specimens in an ice-chamber, temperature 0°C, for 24 hours, and afterwards took their records once more at the ordinary temperature of the room. From these it will be seen that while the responsiveness of Eucharis lily, known to be susceptible to the effect of cold, had entirely disappeared, that of the hardier plants, Holly and Ivy, showed very little change (*Figure 36*).

Another very curious effect that I have noticed is that when a plant approaches its death point by reason of excessively high or low temperature, not only is its general responsiveness diminished almost to zero, but even the slight response occasionally becomes reversed.

Fig. 36 **After-effect of cold on ivy, holly, and eucharis lily**

*a)* The normal response;

*b)* Response after subjection to freezing temperature for 24 hours.

## Influence of high temperature, and determination of death point

I next tried to find out whether a rise of temperature produced a depression of response, and whether the response disappeared at a maximum temperature – the temperature of death point.

For this purpose I took a batch of six radishes and obtained from them responses at gradually increasing temperatures. These specimens were obtained late in the season, and their electric responsiveness was much lower than those obtained earlier. The plant, previously kept for five minutes in water at a definite temperature (say 17°C), was mounted in the vibration apparatus and responses observed.

The plant was then dismounted, and replaced in the water-bath at a higher temperature (say 30°C) again, for five minutes. A second set of responses was now taken. In this way observations were made with each specimen till the temperature at which response almost or altogether ceased was reached. I give below a table of results obtained with six specimens of radish, from which it would appear that response begins to be abolished in these cases at temperatures varying from 53° to 55°C.

## Table showing effect of high temperature in abolishing response

|     | temperature (°C) | Galvanometric response (100 dns. = 0·07 v) |
| --- | --- | --- |
| (1) | 17° | 70 dns |
|     | 53° | 4 dns |
| (2) | 17° | 160 dns |
|     | 53° | 1 dns |
| (3) | 17° | 100 dns |
|     | 50° | 2 dns |
| (4) | 17° | 80 dns |
|     | 55° | 0 dns |
| (5) | 17° | 40 dns |
|     | 60° | 0 dns |
| (6) | 17° | 60 dns |
|     | 55° | 0 dns |

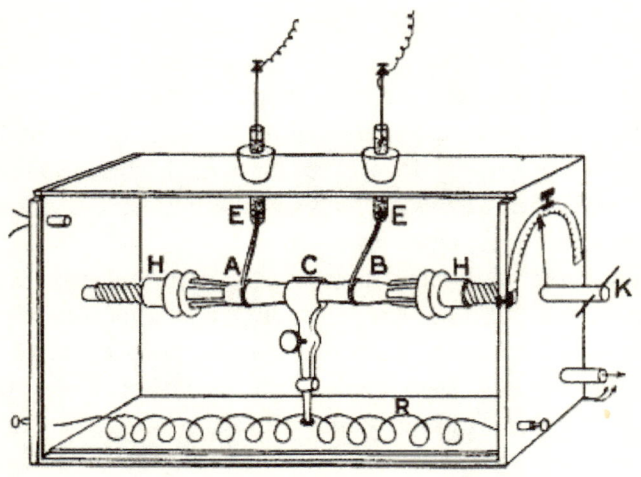

Fig. 37 **The glass chamber containing the plant**

Amplitude of vibration which determines the intensity of stimulus is measured by the graduated circle seen to the right. Temperature is regulated by the electric heating coil **R**. For experiments on the action of anæsthetics, vapour of chloroform is blown in through the side tube.

## Electric heating

The experiments just described were, however, rather troublesome, in that in order to produce each variation of temperature, the specimen had to be taken out of the apparatus, warmed, and remounted. I therefore introduced a modification by which this difficulty was obviated.

The specimen was now enclosed in a glass chamber (*Figure 37*), which also contained a spiral of German silver wire, through which electric currents could be sent, for the purpose of heating the chamber. By varying the intensity of the current, the temperature could be regulated at will. The specimen chosen for experiment was the leaf stalk of celery. It was kept at each given temperature for ten minutes, and two records were taken during that time. It was then raised by 10°C, and the same process was repeated. It will be noticed from the record (*Figure 38*) that in this particular case, as the temperature rose from 20°C to 30°C, there was a marked diminution of response. At the same time, in this case at least, recovery was quicker. At 20°C, for example, the response was 21 dns., and the recovery was not complete in the course of a minute. At 30°C, however, the response had been reduced to 7.5 divisions, but there was almost complete recovery in twelve seconds. As the temperature was gradually increased, a continuous decrease of response occurred.

This diminution of response with increased temperature appears to be universal, but the quickening of recovery may be true of individual cases only.

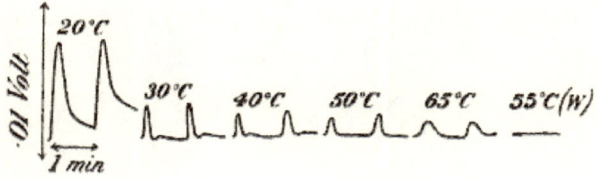

Fig. 38 **Effect of temperature on response**

The response was abolished at the hot-water temperature of 55°C.

## Table showing diminution of response with increasing temperature

(.01 Volt = 35 divisions)

| temperature | response |
|---|---|
| 20° | 21 |
| 30° | 7.5 |
| 40° | 5.5 |
| 50° | 4 |
| 65° | 3 |

In radishes response disappeared completely at 55°C, but with celery, heated in the manner described, I could not obtain its entire abolition at 60°C or even higher. A noticeable circumstance, however, was the prolongation of the period of recovery at these high temperatures. I soon understood the reason of this apparent anomaly. The method adopted in the present case was that of dry heating, whereas the previous experiments had been carried on by the use of hot water. It is well known that one can stand a temperature of 100°C without ill effects in the hot-air chamber of a Turkish bath, while immersion in water at 100°C would be fatal.

In order to find out whether subjection to hot water would kill the celery stalk, I took it out and placed it for five minutes in water at 55°C This, as will be seen from the record taken afterwards, effectively killed the plant (*Figure 38, w*).

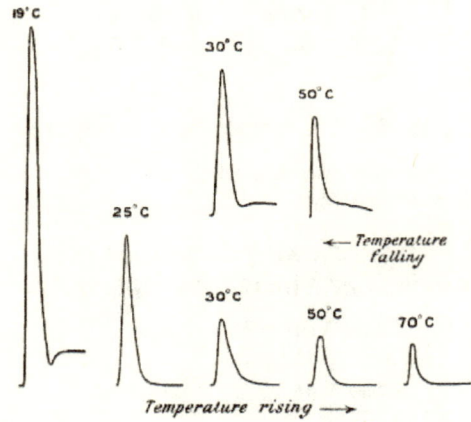

Fig. 39 **Effect of rising and falling temperature on the response of scotch kale**

## Increased sensitiveness as after-effect of temperature variation

A very curious effect of temperature variation is the marked increase of sensitivity which often appears as its after-effect. I noticed this first in a series of observations where records were taken during the rise of temperature and continued while the temperature was falling (*Figure 39*). The temperature was adjusted by electric heating. It was found that the responses were markedly enhanced during cooling, as compared with responses given at the same temperatures while warming (see table). Temperature variation thus seems to have a stimulating effect on response, by increasing molecular mobility in some way. The second record (*Figure 40*) shows the variation of response in Eucharis lily (1) during the rise, and (2) during the fall of temperature. *Figure 41* gives a curve of variation of response during the rise and fall of temperature.

### Table showing the variation of response in scotch kale during the rise and fall of temperature

| temperature | | response (Temperature falling) | | response (Temperature rising) | |
|---|---|---|---|---|---|
| 19°C | | 47 dns | | – | |
| 25°C | ↓ | 24 dns | | – | ↑ |
| 30°C | | 11 dns | | 23 dns | |
| 50°C | | 8 dns | | 16 dns | |
| 70°C | | 7 dns | | – | |
| | | | → | | |

### Fig. 40 **Records of responses in eucharis lily during rise and fall of temperature**

Stimulus constant, applied at intervals of one minute. The temperature of the plant chamber gradually rose on starting current in the heating coil; on breaking current, the temperature fell gradually. Temperature corresponding to each record is given below.

Temperature rising: (1) 20°, (2) 20°, (3) 22°, (4) 38°, (5) 53°, (6) 68°, (7) 65°.

Temperature falling: (8) 60°, (9) 51°, (10) 45°, (11) 40°, (12) 38°.

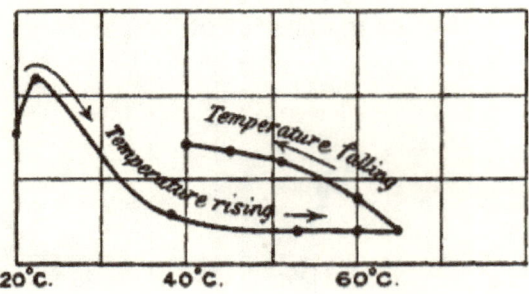

### Fig. 41 **Curve showing variation of response in eucharis with the rise and fall of temperature**

## Point of temperature maximum

We have seen how, in cases of lowered temperature, response is abolished earlier in plants like Eucharis, which are affected by cold, than in the hardier plants such as Holly and Ivy. Plants again are unequally affected as regards the upper range. In the case of Scotch kale, for instance, response disappears after ten minutes of water temperature of about 55°C, but with Eucharis fairly marked response can still be obtained after such immersion and does not disappear till it has been subjected for ten minutes to hot water, at a temperature of 65°C or even higher.

The reason for this great power of resistance to heat is probably found in the fact that the Eucharis is a tropical plant, and is grown, in this country, in hot-houses where a comparatively high temperature is maintained.

## The effect of steam

I next wished to obtain a continuous record by which the effects of suddenly increased temperatures, culminating in the death of the plant, might be made evident. For this purpose I mounted the plant in the glass chamber, into which steam could be introduced. I had chosen a specimen which gave regular response. On the introduction of steam, with the consequent sudden increase of temperature, there was a transitory augmentation of excitability. But this quickly disappeared, and in five minutes the plant was effectively killed, as will be seen graphically illustrated in the record (*Figure 42*).

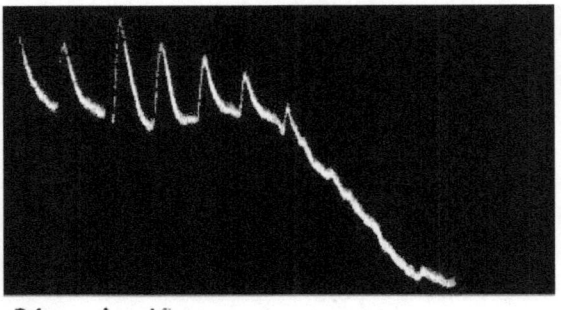

Before ↑ After

### Fig. 42 Effect of steam in killing response

The two readings to the left exhibit normal response at 17°C. Sudden warming by steam produced at first an increase of response, but five minutes exposure to steam killed the plant (carrot) and abolished the response.

Vibrational stimulus of 30° applied at intervals of one minute; vertical line = .1 volt.

It will thus be seen that those modifications of vital activity which are produced in plants by temperature variation can be very accurately gauged by electric response. Indeed it may be said that there is no other method by which the moment of cessation of vitality can be so satisfactorily distinguished.

Ordinarily, we are able to judge that a plant has died, only after various indirect effects of death, such as withering, have begun to appear. But in the electric response we have an immediate indication of the arrest of vitality, and we are thereby able to determine the death point, which it is impossible to do by any other means.

It may be mentioned here that the explanation suggested by Kunkel, of the response being due to movement of water in the plant, is inadequate. For in that case we should expect a definite stimulation to be, under all conditions, followed by a definite electric response, whose intensity and sign should remain invariable.

But we find, instead, the response to be profoundly modified by any influence which affects the vitality of the plant. For instance, the response is at its maximum at an optimum temperature, a rise of a few degrees producing a profound depression; the response disappears at the maximum and minimum temperatures, and is revived when brought back to the optimum. Anæsthetics and poisons abolish the response. Again, we have the response undergoing an actual reversal when the tissue is stale.

All these facts show that mere movement of water could not be the effective cause of plant response.

# Chapter 9
# PLANT RESPONSE – EFFECT OF ANÆSTHETICS AND POISONS

*Effect of anæsthetics, a test of vital character of response – Effect of chloroform – Effect of chloral – Effect of formalin – Method in which response is unaffected by variation of resistance – Advantage of block method – Effect of dose.*


The most important test by which vital phenomena are distinguished is the influence on response of narcotics and poisons. For example, a nerve when narcotised by chloroform exhibits a diminishing response as the action of the anæsthetic proceeds. (See *Figure 43*)

Similarly, various poisons have the effect of permanently abolishing all response. Thus a nerve is killed by strong alkalis and strong acids. I have already shown how plants which previously gave strong response did not, after application of an anæsthetic or poison, give any response at all. In these cases it was the last stage only that could be observed. But it appeared important to be able to trace the growing effect of anæsthetisation or poisoning throughout the process.

There were, however, two conditions which it at first appeared difficult to meet.

First it was necessary to find a specimen which would normally exhibit no fatigue, and give rise for a long time to a uniform series of response. The immediate changes made in the response, in consequence of the application of chemical reagents, could then be demonstrated in a striking manner. And with a little trouble, specimens can be secured in which perfect regularity of response is found. The record given in *Figure 16*, obtained with a specimen of radish, shows how possible it is to secure plants in which response is absolutely regular. I subjected this to uniform stimulation at intervals of one minute, during half an hour, without detecting the least variation in the responses. But it is of course easier to find others in which the responses as a whole may be taken as regular, though there may be slight rhythmic fluctuations. And even in these cases the effect of reagents is too marked and sudden to escape notice.

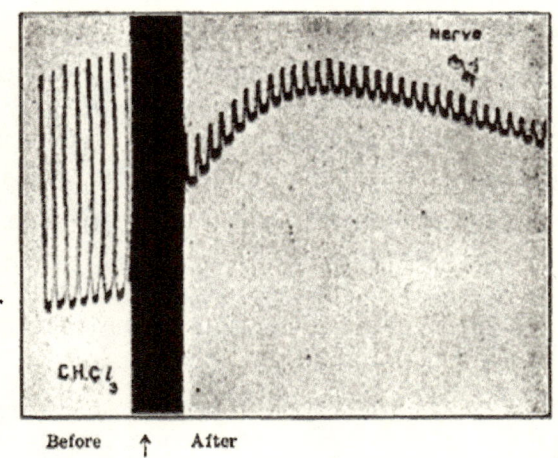

Fig. 43 **Effect of chloroform on nerve response** (Waller)

For the obtaining of constant and strong response I found the best materials to be carrot and radish, selected individuals from which gave most satisfactory results. The carrots were at their best in August and September, after which their sensitiveness rapidly declined.

Later, being obliged to seek other specimens, I came upon radish, which gave good results in the early part of November; but the setting-in of the frost had a prejudicial effect on its responsiveness. Less perfect than these, but still serviceable, are the leaf stalks of turnip and cauliflower. In these the successive responses as a whole may be regarded as regular, though a curious alternation is sometimes noticed, which, however, has a regularity of its own.

My second misgiving was as to whether the action of reagents would be sufficiently rapid to display itself within the time limit of a photographic record. This would of course depend in turn upon the rapidity with which the tissues of the plant could absorb the reagent and be affected by it. It was a surprise to me to find that, with good specimens, the effect was manifested in the course of so short a time as a minute or so.

## The effect of chloroform

In studying the effect of chemical reagents in plants, the method is precisely similar to that employed with nerve; that is to say, where vapour of chloroform is used, it is blown into the plant chamber. In cases of liquid reagents, they are applied on the points of contact A and B and their close neighbourhood.

The mode of experiment was:

(1) to obtain a series of normal responses to uniform stimuli, applied at regular intervals of time, say one minute, the record being taken on a photographic plate.

(2) Without interrupting this procedure, the anæsthetic agent (vapour of chloroform) was blown into the closed chamber containing the plant. It will be seen how rapidly chloroform produces depression of response (*Figure 44*), and how the effect grows with time.

In these experiments with plants, the same curious shifting of the zero line is sometimes noticed as in nerve when subjected similarly to the action of reagents. This is a point of minor importance; the essential point is that the responses are rapidly reduced.

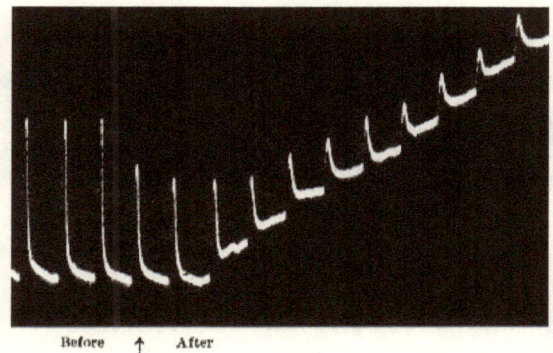

Before ↑ After

### Fig. 44 Effect of chloroform on responses of carrot

Stimuli of 25° vibration at intervals of one minute.

## Effects of chloral and formalin

I give below (*Figures 45, 46*) two sets of records, one for the reagent chloral and the other for formalin. The reagents were applied in the form of a solution on the tissue at the two leading contacts, and the contiguous surface. The rhythmic fluctuation in the normal response shown in *Figure 45* is interesting. The abrupt decline, within a minute of the application of chloral, is also extremely well marked.

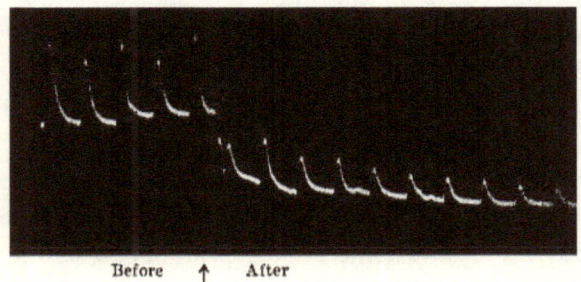

Before ↑ After

### Fig. 45 Action of chloral hydrate on the responses of leaf stalk of cauliflower

Vibration of 25° at intervals of one minute.

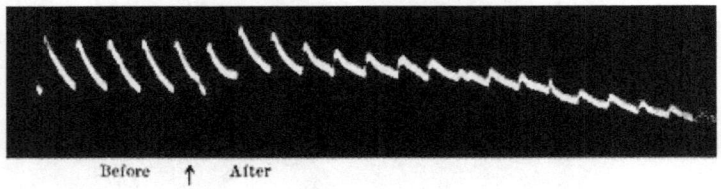

Fig. 46 **Action of formalin (radish)**

## Response unaffected by variation of resistance

In order to bring out clearly the main phenomena, I have postponed till now the consideration of a point of some difficulty.

To determine the influence of a reagent in modifying the excitability of the tissue, we rely upon its effect in exalting or depressing the responsive E.M. variation. We read this effect by means of galvanometric deflections. And if the resistance of the circuit remained constant, then an increase of galvanometer deflection would accurately indicate a heightened or depressed E.M. response, due to greater or less excitability of tissue caused by the reagent. But, with the introduction of the chemical reagent, the resistance of the tissue may undergo change, and owing to this cause, modification of response as read by the galvanometer may be produced without any E.M. variation. The observed variation of response may thus be partly owing to some unknown change of resistance, as well as to that of the E.M. variation in response to stimulus.

We may however establish how much of the observed change is due to variation of resistance by comparing the deflections produced in the galvanometer by the action of a definite small E.M.F. before and after the introduction of the reagent. If the deflections are the same in both cases, we know that the resistance has not varied. If there has been any change, the variation of deflection will show the amount, and we can make allowance accordingly.

I have however adopted another method, in which all necessity of correction is obviated, and the galvanometric deflections simply give E.M. variations, unaffected by any change in the resistance of the tissue.

This is done by interposing a very large and constant resistance in the external circuit and thereby making other resistances negligible. An example will make this point clear.

Taking a carrot as the vegetable tissue, I found its resistance plus the resistance of the non-polarisable electrode equal to 20,000 ohms. The introduction of a chemical reagent reduced it to 19,000 ohms. The resistance of the galvanometer is equal to 1,000 ohms. The high external resistance was 1,000,000 ohms. The variation of resistance produced in the circuit would therefore be 1,000 in (1,000,000 + 19,000 + 1,000) or one part in 1,020. Therefore the variation of galvanometric deflection due to change of resistance would be less than one part in a thousand (*cf. Figure 49*).

## The advantage of the block method

In these investigations I have used the block method, instead of that of negative variation, and I may here draw attention to the advantages which it offers.

In the method of negative variation, one contact being injured, the chemical reagents act on injured and uninjured unequally, and it is conceivable that by this unequal action the resting difference of potential may be altered. But the intensity of response in the method of injury depends on this resting difference. It is thus hypothetically possible that with the method of negative variation there might be changes in the responses caused by variation of the resting difference, and not necessarily due to the stimulating or depressing effect of the reagent on the tissue.

But by the block method the two contacts are made with uninjured surfaces, and the effect of reagents on both is similar. Thus no advantage is given to one contact over the other. The changes now detected in response are therefore due to no adventitious circumstance, but to the reagent itself.

If further verification be desired as to the effect of the reagent, we can obtain it through alternate stimulation of the A and B ends. Both ends will then show the given change. I give below a record of responses given by two ends of leaf stalk of turnip, stimulated alternately in the manner described. The stalk used was slightly conical, and owing to this difference between the A and B ends the responses given by one end were slightly different from those given by the other, though the stimuli were equal. A few drops of a 10% solution of NaOH was applied to both the ends. It will be seen how quickly this reagent abolished the response of both ends (*Figure 47*).

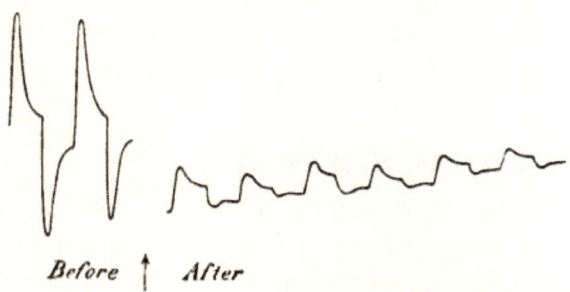

*Before* ↑ *After*

### Fig. 47 Abolition of response at both A and B ends by the action of NaOH

Stimuli of 30° vibration were applied at intervals of one minute to **A** and **B** alternately. Response was completely abolished 24 minutes after the application of NaOH.

## Effect of dose

It is sometimes found that while a reagent acts as a poison when given in large quantities, it may act as a stimulant in small doses. Of the two following records *Figure 48* shows the slight stimulating effect of very dilute KOH, and *Figure 49* exhibits nearly complete abolition of response by the action of the same reagent when given in stronger doses.

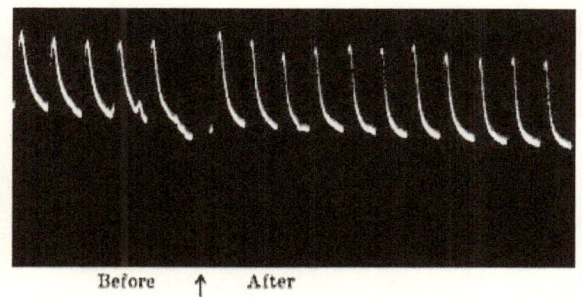

Fig. 48 **Stimulating action of very dilute KOH**

So we see that, judged by the final criterion of the effect produced by anæsthetics and poisons, the plant response fulfils the test of vital phenomenon. In previous chapters we have found that in the matter of response by negative variation, of the presence or absence of fatigue, of the relation between stimulus and response, of modification of response by high and low temperatures, and even in the matter of occasional abnormal variations such as positive response in a modified tissue, *they were strictly correspondent to similar phenomena in animal tissues.* The remaining test, of the influence of chemical reagents, having now been applied, a complete parallelism may be held to have been established between plant response on the one hand, and that of animal tissue on the other.

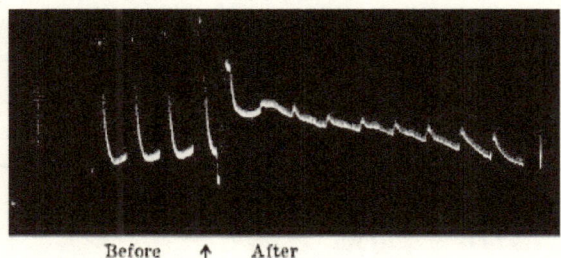

Fig. 49 **Nearly complete abolition of response by strong KOH**

The two vertical lines are galvanometer deflections due to ·1 volt, before and after the application of reagent. It will be noticed that the total resistance remains unchanged.

# Chapter 10
# RESPONSE IN METALS

*Is response found in inorganic substances?*
*– Experiment on tin, block method – Anomalies*
*of existing terminology – Response by method of*
*depression – Response by method of exaltation.*

We have now seen that electrical signs of life are not confined to animals, but is also found in plants. And we have seen how electrical response serves as an index to the vital activity of the plant, how with the arrest of this vital activity electrical response is also arrested temporarily, as in the case amongst others of anæsthetic action, and permanently, for instance under the action of poisons. Thus living tissues – both animal and vegetable – may pass from a responsive to an irresponsive condition, from which latter there may or may not be subsequent revival.

Hitherto, as already said, electrical response in animals has been regarded as a purely physiological phenomenon. We have proved by various tests that response in plants is of the same character. And we have seen that by physiological phenomena are generally understood those of which no physical explanation can be offered, they being supposed to be due to the play of some unknown vital force existing in living substances and giving rise to electric response to stimulation as one of its manifestations.

## Is response found in inorganic substances?[14]

It is now for us, however, to examine the alleged super-physical character of these phenomena by stimulating inorganic substances and discovering whether they do or do not give rise to the same electrical mode of response which was supposed to be the special characteristic of living substances. We shall use the same apparatus and the same mode of stimulation as those employed in obtaining plant response, merely substituting, for the stalk of a plant, a metallic wire, say 'tin' (*Figure 50*). Any other metal could be used instead of tin.

## Experiment on tin, block method

Let us then take a piece of tin wire[15] from which all strains have been previously removed by annealing, and hold it clamped in the middle at C. If the strains have been successfully removed A and B will be found to be iso-electric, and no current will pass through the galvanometer. If A and B are *not* exactly similar, there will be a slight current. But this will not materially affect the results to be described presently, the slight existing current merely adding itself algebraically to the current of response.

If we now stimulate the end A by taps, or better still by torsional vibration, a transitory 'current of action' will be found to flow in the wire from B to A, from the unstimulated to the stimulated, and in the galvanometer from the stimulated to the unstimulated. Stimulation of B will give rise to a current in an opposite direction.

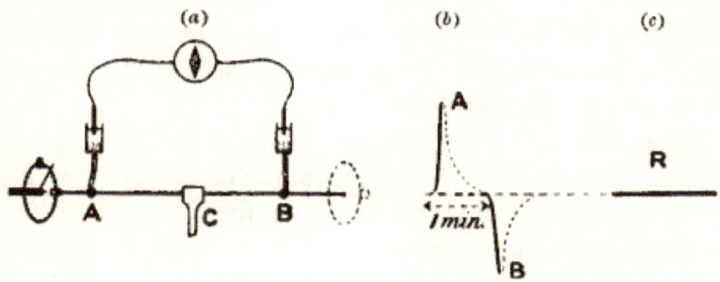

Fig. 50 **Electric response in metals**

*(a)* Method of block; *(b)* Equal and opposite responses when the ends **A** and **B** are stimulated; the dotted portions of the curves show recovery; *(c)* Balancing effect when both the ends are stimulated simultaneously.

## Experiment to exhibit the balancing effect

If the wire has been carefully annealed, the molecular condition of its different portions is found to be approximately the same. If such a wire be held at the 'balancing point' (which is at or near the middle) by the clamp, and a quick vibration, say, of 90° is given to A, an upward deflection will be produced; if a vibration of 90° is given to B, there will be an equal downward deflection. If now both the ends A and B are vibrated simultaneously, the responsive E.M. variation at the two ends will continuously balance each other and the galvanometer spot will remain quiescent (*Figure 30*, A, B, R). This balance will be still maintained when the block is removed and the wire is vibrated as a whole.

It is to be remembered that with the length of wire constant, the intensity of stimulus increases with the amplitude of vibration.

Again, keeping the amplitude constant, the intensity of stimulus is increased by shortening the wire. Hence it will be seen that if the clamp is shifted from the balancing point towards A, simultaneous vibration of A and B through 90° will now give a resultant upward deflection, showing that the A response is now relatively stronger. Thus keeping the rest of the circuit untouched, merely moving the clamp from the left, past the balancing point to the right, we get either a positive, or zero, or negative, resultant effect.

In tin the current of response is from the less to the more excited point. In the retina also, we found the current of action flowing from the less stimulated to the more stimulated, and as that is known as a positive response, we shall consider the normal response of tin to be in like manner positive.

Just as the response of retina or nerve, under certain molecular conditions, undergoes reversal, the positive being then converted into negative, and negative into positive, so it will be shown that the response in metallic wires under certain conditions is found to undergo reversal.

## Anomalies of present terminology

When there is no current of injury, a particular current of response can hardly be called a negative, or positive, *variation*. Such nomenclature is purely arbitrary, and leads, as will be shown, to much confusion. A more definite terminology, free from misunderstanding, would be, as already said, to regard the current towards the more stimulated as *positive*, and that towards the less stimulated, in tissue or wire, as *negative*.

The stimulated end of tin, say the end A, thus becomes *zincoid*, i.e. the current through the electrolyte (non-polarisable electrodes with interposed galvanometer) is from A to B, and *through the wire*, from the less stimulated B to the more stimulated A. Conversely, when B is stimulated, the action current flows round the circuit in an opposite direction. This positive is the most usual form of response, but there are cases where the response is negative.

In order to show that normally speaking a stimulated wire becomes zincoid, and also to show once more the anomalies into which we may fall by adopting no more definite terminology than that of negative variation, I have devised the following experiment (*Figure 51*).

Let us take a bar, one half of which is zinc and the other half copper, clamped in the middle, so that a disturbance produced at one end may not reach the other; the two ends are connected to a galvanometer through non-polarisable electrodes. The current through the electrolyte (non-polarisable electrodes and interposed galvanometer) will then flow from left to right. We must remember that metals under stimulation generally become, in an electrical sense, more zinc-like. On vibrating the copper end (inasmuch as copper would then become more zinc-like) the difference of potential between zinc and copper ought to be diminished, and the current flowing in the circuit would therefore be lessened. But vibration of the zinc end ought to increase the potential difference, and there ought to be then an increase of current during stimulation of zinc.

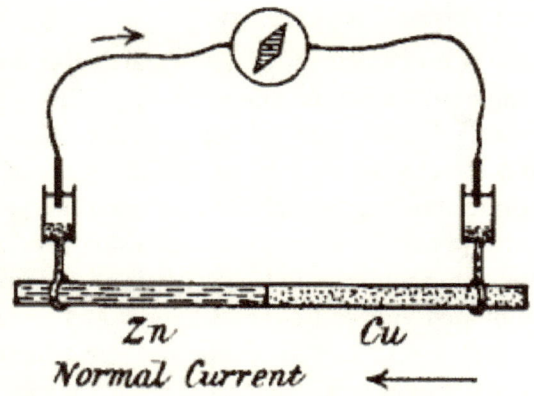

### Fig. 51 **Current of response towards the stimulated end**

Hence when Cu is stimulated: action current →, normal E.M.F. diminished (.85 − .009) V.

When Zn stimulated: action current ←, normal E.M.F. increased (.85 + .013) V.

In the particular experiment of *Figure 51*, the E.M.F. between the zinc and copper ends was found to be .85 volt. This was balanced by a potentiometer arrangement, so that the galvanometer spot came to zero. On vibrating the zinc wire, a deflection of 33 dns. was obtained, in a direction which showed an *increase* of E.M.F. On stopping the vibration, the spot of light came back to zero. On now vibrating the copper wire, a deflection of 23 dns. was obtained in an opposite direction, showing a *diminution* of E.M.F. This transitory responsive variation disappeared on the cessation of disturbance.

By disturbing the balance of the potentiometer, the galvanometer deflection due to a known increase of E.M.F. was found from which the absolute E.M. variation caused by disturbance of copper or zinc was determined.

It was thus found that stimulation of zinc had *increased* the P.D. by fifteen parts in 1,000, whereas stimulation of copper had *decreased* it by eleven parts in 1,000. According to the old terminology, the response due to stimulation of zinc would have been regarded as positive variation, that of copper negative. The responses however are not essentially opposite in character, the action current in the bar being in both cases towards the more excited. For this reason it would be preferable, as already said, to employ the terms *positive* and *negative* in the sense I have suggested, i.e. *positive*, when the current in the acted substance is towards the more excited, and *negative*, when towards the less excited. The method of block is, as I have already shown, the most suited for the study of these responses.

In the experiment in *Figure 50*, if the block is abolished and the wire is struck in the middle, a wave of molecular disturbance will reach A and B. The mechanical and the attendant electrical disturbance will at these points reach a maximum and then gradually subside. The resultant effect in the galvanometer will be due to $E_A - E_B$ when $E_A$ and $E_B$ are the electrical variations produced at A and B by the stimulus. The electric changes at A and B will continuously balance each other, and the resultant effect on the galvanometer will be zero:

(*a*) if the exciting disturbance reaches A and B at the same time and with the same intensity;

(*b*) if the molecular condition is similar at the two points; and (*c*) if the rate of rise and subsidence of excitation is the same at the two points. In order that a resultant effect may be exhibited in the galvanometer, matters have to be so arranged that the disturbance may reach one point, say A, and not B, and *vice versa*. This was accomplished by means of a clamp, in the method of block. Again a resultant differential action may be obtained even when the disturbance reaches both A and B, if the electrical excitability of one point is exalted or depressed by physical or chemical means.

In *Chapter 16* we shall study in detail the effect of chemical reagents in producing the enhancement or depression of excitability.

There are thus two other means of obtaining a resultant

effect – (2) by the method of relative depression, (3) by the method of relative exaltation.

## Electric response by method of depression

We may thus by reducing or abolishing the excitability of one end by means of suitable chemical reagents (so-called method of injury) obtain response in metals without a block. The entire length of the wire may then be stimulated and a resultant response will be produced, owing to the difference between the excitability of the two ends.

A piece of tin wire is taken, and one normal contact is made at A (with a strip of cloth moistened with water, or very dilute salt solution). The excitability of B is depressed by a few drops of strong potash or oxalic acid. By the application of the latter there will be a small P.D. between A and B; this will simply produce a displacement of zero.

By means of a potentiometer the galvanometer spot may be brought back to the original position. The shifting of the zero will not affect the general result.

The effect of mechanical stimulus is to produce a transient electro-motive response, which will be superposed algebraically on the existing P.D. The deflection will take place from the modified zero to which the spot returns during recovery. On now stimulating the wire as a whole by, say, torsional vibration, the current of response will be found towards the more excitable, i.e. from B to A (*Figure 52a*).

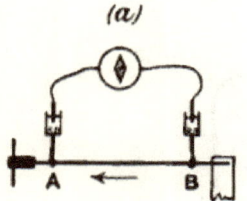 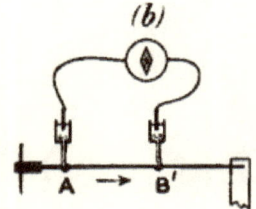

**Fig. 52 Response by method of depression (without block)**

When the wire is stimulated as a whole the current of response is towards the more excitable.

In *(a)* A is a normal contact, B has been depressed by oxalic acid; current of response is towards the more excitable A.

In *(b)* the same wire is used, but A is depressed by oxalic acid and a normal contact is made at a fresh point B′, a little to the left of B in (a). The current of response is now from A towards the more excitable B′.

A corroborative reversal experiment may next be made on the same piece of wire. The normal contact, through water or salt solution, is now made at B′, a little to the left of B. The excitability of A is now depressed by oxalic acid. On stimulation of the whole wire, the current of response will now be found to flow in an opposite direction – i.e. from A to B′ – but still from the relatively less to the relatively more excitable (*Figure 52b*).

From these experiments it can be seen how in one identical piece of wire the responsive current flows now in one direction and then in the other, in absolute conformity with theoretical considerations.

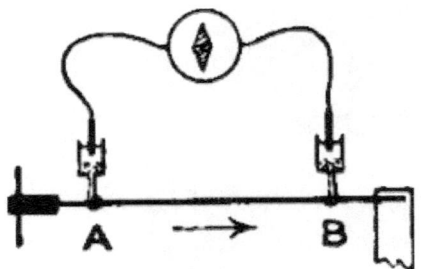

Fig. 53 **Method of exaltation**

The contact **B** is made more excitable by chemical stimulant ($Na_2CO_3$). The current of response is towards the more excitable **B**.

## Method of exaltation

A still more striking corroboration of these results may, however, be obtained by the converse process of relative exaltation of the responsiveness of one contact. This may be accomplished by touching one contact, say B, with a reagent which like $Na_2CO_3$ increases the electric excitability. On stimulation of the wire, the current of response is towards the more excitable B (*Figure 53*).

I give four records (*Figure 54*) which will clearly exhibit the responses as obtained by the methods of relative depression or exaltation.

In (*a*) B is touched with the excitant $Na_2CO_3$. A permanent current flows from A to B, response to stimulus is in the same direction as the permanent current (positive variation).

In (*b*) B is touched with a trace of the depressant oxalic acid. The permanent current is in the same direction as before, but the current of response is in the opposite direction (negative variation).

In (*c*) B is touched with dilute KHO. The response is exhibited by a positive variation.

In (*d*) B is touched with strong KHO. The response is now

exhibited by a negative variation.

The last two results, apparently anomalous, are due to the fact, which will be demonstrated later, that KHO in minute quantities is an excitant, while in large quantities it is a depressant.

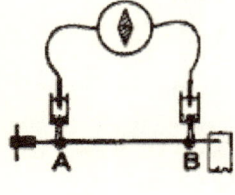 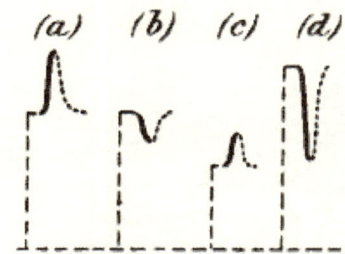

Fig. 54

|  | Permanent Current | Current of response |
|---|---|---|
| **B** treated with *sodium carbonate* | → | → |
| **B** treated with *oxalic acid* | → | ← |
| **B** treated with *very dilute potash* | → | → |
| **B** treated with *strong potash* | → | ← |

Current of response is always towards the more excitable point.

*a)* Response when **B** is treated with sodium carbonate.
   – An apparent positive variation.

*b)* Response when **B** is treated with oxalic acid.
   – An apparent negative variation.

*c)* Response when **B** is treated with very dilute potash.
   – Positive variation.

*d)* Response when **B** is treated with strong potash.
   – Negative variation.

The response is up when **B** is more excitable,
and down when **A** is more excitable.

Lines thus ------ indicate deflection due to permanent current.

We have thus seen that we may obtain response (1) by block method, (2) by the method of injury, or relative depression of responsiveness of one contact, and (3) by the method of relative exaltation of responsiveness of one contact. In all these cases alike we obtain a consistent action current, which in tin is normally positive, or towards the relatively more excited.

## Footnotes

14. Following another line of inquiry I obtained response to electric stimulus in inorganic substances using the method of conductivity variation (see *De la Généralité des Phénomènes Moléculaires Produits par l'Electricité sur la Matière Inorganique et sur la Matière Vivante*, Travaux du Congrès International de Physique, Paris, 1900; and also *On Similarities of Effect of Electric Stimulus on Inorganic and Living Substances*, British Association 1900. See *Electrician*). To illustrate the parallels in all details between the inorganic and living response, I have in the following chapters used the method of electro-motive variation employed by physiologists.

15. By 'tin' is meant an alloy of tin and lead, used as electric fuse.

# Chapter 11

# INORGANIC RESPONSE – MODIFIED APPARATUS TO EXHIBIT RESPONSE IN METALS

*Conditions of obtaining quantitative measurements – Modification of the block method – Vibration cell – Application of stimulus – Graduation of the intensity of stimulus – Considerations showing that electric response is due to molecular disturbance – Test experiment – Molecular voltaic cell.*

We have already seen that metals respond to stimulus by E.M. variation, just as do animal and vegetable tissues. We have yet to see whether the similarity extends to this point only, or goes still further – whether the response curves of living and in organic are alike, and whether the inorganic response curve is modified, as living response was found to be, by the influence of external agencies.

If so, are the modifications similar? What are the effects of superposition of stimuli? Is there fatigue? If there is, in what way does it affect the curves? And lastly, is the response of metals increased or lessened by the action of chemical reagents?

## Conditions of obtaining quantitative measurements

In order to carry out these investigations, it is necessary to remove all sources of uncertainty, and obtain quantitative measurements. Many difficulties at first presented themselves in the course of this, but they were completely removed by the adoption of the following experimental modification.

In the simple arrangement for qualitative demonstration of

response in metals previously described, successive experiments will not give results which are strictly comparable;

(1) unless the resistance of the circuit is maintained at a constant. This would necessitate the adoption of some plan for keeping the electrolytic contacts at A and B absolutely invariable. There should then be no chance of any shifting or variation of contact.

(2) There must also be some means of applying successive stimuli of equal intensity.

(3) And for certain further experiments it will be necessary to have some way of gradually increasing or decreasing the stimuli in a definite manner.

## Modification of the block method

By consideration of the following experimental modifications of the block method (*Figure 55*), it will be easy to construct a perfected form of apparatus, in which all these conditions are fully met. The essentials to be kept in mind are the introduction of a complete block midway in the wire, so that the disturbance of one half should be prevented from reaching the other, and the making of a perfect electrolytic contact for the electrodes leading to the galvanometer.

Starting from the simple arrangement previously described where a straight wire is clamped in the middle (*Figure 55a*), we next arrive at (*b*). Here the wire A B is placed in a U tube and clamped in the middle by a tightly fitting cork. Melted paraffin wax is poured to a certain depth in the bend of the tube. The two limbs of the tube are now filled with water, till the ends A and B are completely immersed. Connection is made with the non-polarisable electrodes by the side tubes. Vibration may be imparted to either A or B by means of ebonite clip holders seen at the upper ends A B of the wire.

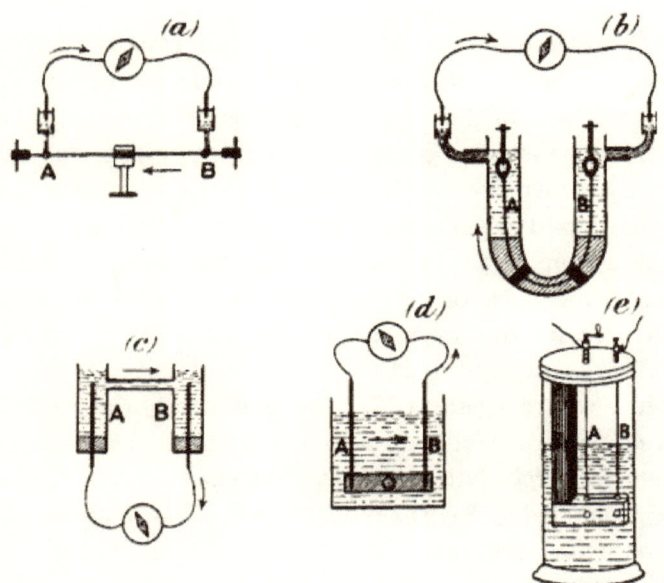

**Fig. 55 Successive modifications of the block method from the 'straight wire' (a) to 'cell form' (e)**

When A is excited, current of response in the wire is from less excited B to more excited A. Note that though the current of response is constant in direction, the galvanometer deflection in *(d)* will be opposite to that in *(b)*.

It will be seen that the two limbs of the tube filled with water serve the purpose of the strip of moistened cloth used in the last experiment to make electric connections with the leading-out electrodes – with the advantage that we have here no chance of any shifting of contact or variation of surface, the contact between the wire and the surrounding liquid being perfect and invariable.

On now vibrating the end A of the tin wire by means of the ebonite clip holder, a current will be found to flow from B to A through the wire – that is to say, towards the excited – and from A to B in the galvanometer.

The next modification (*c*) is to transfer the galvanometer

from the electrolytic to the metallic part of the circuit, that is to say, it is interposed in a gap made by cutting the wire A B, the upper part of the circuit being directly connected by the electrolyte. Vibration of A will now give rise to a current of response which flows in the metallic part of the circuit with the interposed galvanometer from B to A. We see that though the direction of the current in this is the same as in the last case, yet the galvanometer deflection is now reversed, for the evident reason that we have it interposed in the metallic and not in the electrolytic part of the circuit.

The next arrangement (*d*) consists simply of the preceding placed upside down. Here A and B are held parallel to each other in an electrolytic bath (water). Mechanical vibration may now be applied to A without affecting B, and *vice versa*.

The actual apparatus, of which this is a diagrammatic representation, is seen in (*e*).

Two pieces, from the same specimen of wire, are clamped separately at their lower ends by means of ebonite screws, in an L-shaped piece of ebonite. The wires are fixed at their upper ends to two electrodes – leading to the galvanometer – and kept moderately and uniformly stretched by spiral springs. The handle, by which a torsional vibration is imparted to the wire, may be slipped over either electrode. The amplitude of vibration is measured by means of a graduated circle.

It will be seen from these arrangements:

(1) That the cell depicted in (*e*) is essentially the same as that in (*a*).

(2) That when the wires in the cell are immersed to a definite depth in the electrolyte, there is always a perfect and invariable contact between the wire and the electrolyte. The difficulty as regards variation of contact is thus eliminated.

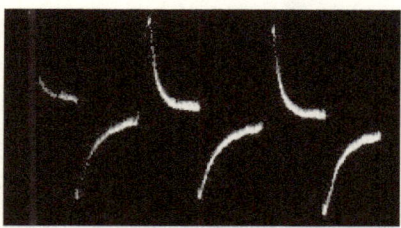

Fig. 56 **Equal and opposite responses exhibited by A and B**

(3) That as the wires A and B are clamped separately below, we may impart a sudden molecular disturbance to either A or B by giving a quick to-and-fro (torsional) vibration round the vertical wire, as axis, by means of the handle. As the wire A is separate from B, disturbance of one will not affect the other. Vibration of A produces a current in one direction, vibration of B in the opposite direction. Thus we have means of verifying every experiment by obtaining corroborative and reversed effects. When the two wires have been brought to exactly the same molecular condition by the processes of annealing or stretching, the effects obtained on subjecting A or B to any given stimulus are always equal (*Figure 56*).

Usually I interpose an external resistance varying from one to five megohms according to the sensitiveness of the wire. The resistance of the electrolyte in the cell is thus relatively small, and the galvanometer deflections are proportional to the E.M. variations. It is always advisable to have a high external resistance, as by this means one is not only able to keep the deflections within the scale, but one is not troubled by slight accidental disturbances.

## Graduation of intensity of stimulus

If now a rapid torsional vibration is given to A or B, an E.M. variation will be induced. If the amplitude of vibration is kept constant, successive responses – in substances which, like tin, show no fatigue – will be found to be absolutely identical. But as 'the amplitude of vibration' is increased, response will also become enhanced (see *Chapter 15*).

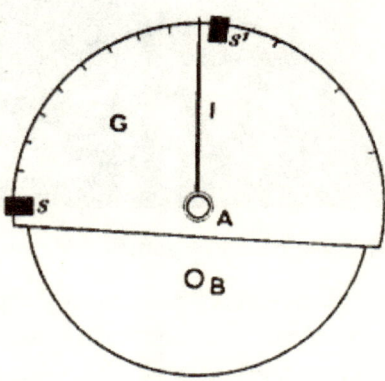

Fig. 57 **Top view of the vibration cell**

The amplitude of vibration is determined by means of movable stops S and S', fixed to the edge of the graduated circle G. The index arm I plays between the stops. (The second index arm, connected with B, and the second circle are not shown.)

Amplitude of vibration is measured by means of the graduated circle (*Figure 57*). A projecting index, in connection with the vibration head, plays between the fixed and sliding stops (S and S'), one at the zero point of the scale, and the other movable. The amplitude of a given vibration can thus be predetermined by the adjustment of the sliding stop. In this way we can obtain either uniform or definitely graduated stimuli.

## Considerations showing that electric response is due to molecular disturbance

The electromotive variation varies with the substance. With superposition of stimuli, a relatively high value is obtained in tin, amounting sometimes to nearly half a volt, whereas in silver the electromotive variation is only about ·01 of this value. The intensity of the response, however, does not depend on the chemical activity of the substance, for the electromotive variation in the relatively chemically inactive tin is greater than that of zinc. Again, the sign of response, positive or negative, is sometimes modified by the molecular condition of the wire (see *Chapter 12*).

As regards the electrolyte, dilute NaCl solution, dilute solution of bichromate of potash, etc, are normal in their action, that is to say, the electric response in such electrolytes is practically the same as with water. Ordinarily I use tap water as the electrolyte. Zinc wires in $ZnSO_4$ solution give responses similar in character to those given by, for example, Pt or Sn in water.

## Test experiment

It may be suggested that the E.M. effect is due in some way

(1) to the friction of the vibrating wire against the liquid; or

(2) to some unknown surface action, at the point in the wire of the contact of liquid and air surfaces. This second objection has already been completely met in experimental modification, *Figure 55b*, where the wire was shown to give response when kept completely immersed in water, variation of surface being thus entirely eliminated.

Both these questions may, however, be subjected to a definite and final test. When the wire to be acted on is clamped below, and vibration is imparted to it, a strong molecular disturbance is produced. If it is then carefully released from the clamp, and the wire rotated backwards and forwards, there could be little molecular disturbance, but the liquid friction and surface variation, if any, would remain. The effect of any slight disturbance outstanding owing to shaking of the wire would be relatively very small.

We can thus determine the effect of liquid friction and surface action by repeating an experiment with and without clamping. In a tin wire cell, with interposed external resistance equal to one million ohms, wire A was subjected to a series of vibrations through 180°, and a deflection of 210 divisions was obtained. A corresponding negative deflection resulted on vibrating wire B. Now A was released from the clamp, so that it could be rotated backwards and forwards in the water by means of the handle. On vibrating wire A no measurable deflection was produced, thus showing that neither water friction nor surface variation had anything to do with the electric action. The vibration of the still clamped B gave rise to the normal strong deflection.

As all the rest of the circuit was kept absolutely the same in the two different sets of experiments, these results conclusively prove that the responsive electro-motive variation is solely due to the molecular disturbance produced by mechanical vibration in the acted wire.

A new and theoretically interesting molecular voltaic cell may thus be made, in which the two elements consist of the *same metal*. Molecular disturbance is in this case the main source of energy. A cell once made may be kept in working order for some time by pouring in a little vaseline to prevent evaporation of the liquid.

It will be shown further, in succeeding chapters, by numerous instances, that any conditions which increase molecular mobility will also increase intensity of response, and conversely that any conditions having the reverse effect will depress response.

# Chapter 12
# INORGANIC RESPONSE – METHODS OF ENSURING CONSISTENT RESULTS

*Preparation of wire – Effect of single stimulus.*

I shall now proceed to describe in detail the response curves obtained with metals. The E.M. variations resulting from stimulus range, as has been said, from .4 volt to .01 of that value, depending on the metal employed. And as these are molecular phenomena, the effect will also depend on the molecular condition of the wire.

## Preparation of wire

In order to make sure that our results are thoroughly consistent, it is necessary to bring the wire itself into a normal condition for experiment. The very fact of mounting it in the cell strains it, and the after-effect of this strain may cause irregularities in the response.

For the purpose of bringing the wire to this normal state, one or all of the following devices may be used.

(1) The wires obtained are usually wound on spools. It is, therefore, advisable to straighten any given length, before mounting, by holding it stretched, and rubbing it up and down with a piece of cloth. On washing with water, they are now ready for mounting in the cell.

(2) The cell is usually filled with tap water, and a period of rest after making up, generally speaking, improves the sensitiveness. These expedients are ordinarily sufficient, but it occasionally happens that the wire has got into an abnormal condition.

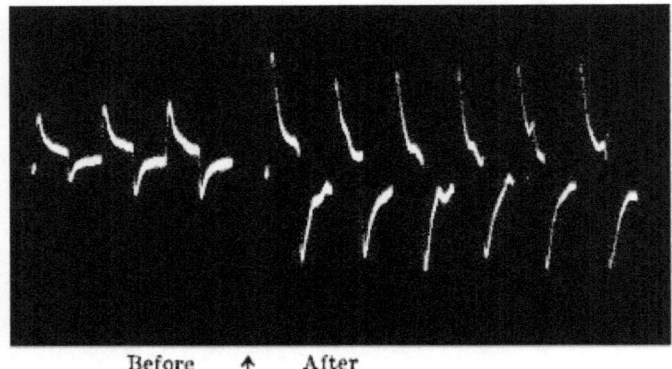

Before ↑ After

### Fig. 58 **Effect of annealing on increasing the response of both A and B wires (tin)**

Stimuli (vibration of 160°) applied at intervals of one minute.

In this case it will be found helpful;

(3) to have recourse to the process of annealing. For if response is a molecular phenomenon, then anything that increases molecular mobility will also increase its intensity. Hence we may expect annealing to enhance responsiveness. This inference will be seen verified in the record given in *Figure 58*. In the case under consideration, the convenient method employed was by pouring hot water into the cell, and allowing it to stand and cool slowly. The first three pairs of responses were taken by stimulating A and B alternately, on mounting in the cell, which was filled with water. Hot water was then substituted, and the cell was allowed to cool down to its original temperature. The six following pairs of responses were then taken. That this beneficial effect of annealing was not due to any accidental circumstance will be seen from the fact that *both* wires have their sensitiveness equally enhanced.

(4) In addition to this mode of annealing, both wires may be short-circuited and vibrated for a time.

(5) Lastly slight stretching *in situ* will also sometimes be found to be beneficial. For this purpose I have a screw arrangement.

By one or all of these methods, with a little practice, it is always possible to bring the wires to a normal condition. The responses subsequently obtained become extraordinarily consistent. There is therefore no reason why perfect results should not be arrived at.

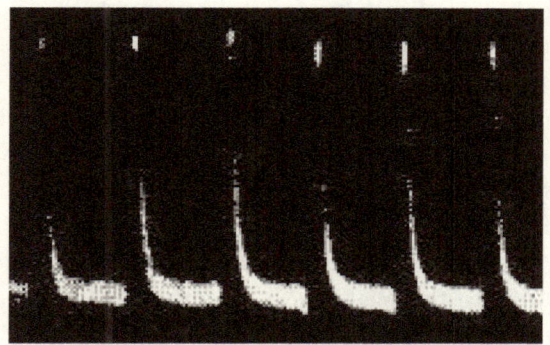

Fig. 59 **Uniform responses in tin**

## Effect of single stimulus

The above figure (*Figure 59*) gives a series, each of which is the response curve for a single stimulus of uniform intensity, the amplitude of vibration being kept constant. The perfect regularity of responses will be noticed in this figure.

The wire after a long period of rest may be in an abnormal condition, but after a short period of stimulation the responses become extremely regular, as may be noticed in this figure. Tin is, usually speaking, almost indefatigable, and I have often obtained several hundreds of successive responses showing practically no fatigue.

In the figure it will be noticed that the rising portion of the curve is somewhat steep, and the recovery convex to the abscissa, the fall being relatively rapid in its first, and less rapid in its later, parts. As the electric variation is the concomitant effect of molecular disturbance – a temporary upset of the molecular equilibrium – on the cessation of the external stimulus, the excitatory state, and its expression in electric variation, disappear with the return of the molecules to their condition of equilibrium. This process is seen clearly in the curve of recovery.

Different metals exhibit different periods of recovery, and this again is modified by any influence which affects the molecular condition.

That the excitatory state persists for a time even on the cessation of stimulus can be independently shown by keeping the galvanometer circuit open during the application of stimulus, and completing it at various short intervals after the cessation, when a persisting electrical effect, diminishing rapidly with time, will be apparent. The rate of recovery immediately on the cessation of stimulus is rather rapid, but traces of strain persist for a short time.

# Chapter 13
# INORGANIC RESPONSE – MOLECULAR MOBILITY: ITS INFLUENCE ON RESPONSE

*Effects of molecular inertia – Prolongation of period of recovery by overstrain – Molecular model – Reduction of molecular sluggishness attended by quickened recovery and heightened response – Effect of temperature – Modification of latent period and period of recovery by the action of chemical reagents – Diphasic variation.*

We have seen that the stimulation of matter causes an electric variation, and that the material being tested gradually recovers from the effect of stimulus. We shall next study how the form of response curves is modified by various agencies.

In order to study these effects we must use, in practice, a highly sensitive galvanometer as the recorder of E.M. variations. This necessitates the use of an instrument with a comparatively long period of swing of needle, or of suspended coil (as in a D'Arsonval). Owing to the inertia of the recording galvanometer, however, there is a lag produced in the records of E.M. changes. But this can be distinguished from the effect of the molecular inertia of the substance itself by comparing two successive records taken with the same instrument, in one of which the latter effect is relatively absent, and in the other present. We wish, for example, to find out whether the E.M. effect of mechanical stimulus is instantaneous, or, again, whether the effect disappears immediately. We first take a galvanometer record of the sudden introduction and cessation of an E.M.F. on the circuit containing the vibration cell (*Figure*

60a). We then take a record of the E.M. effect produced by a stimulus caused by a single torsional vibration. In order to make the conditions of the two experiments as similar as possible, the disturbing E.M.F., from a potentiometer, is previously adjusted to give a deflection nearly equal to that caused by stimulus. The torsional vibration was accomplished in a quarter of a second, and the contact with the potentiometer circuit was also made for the same length of time.

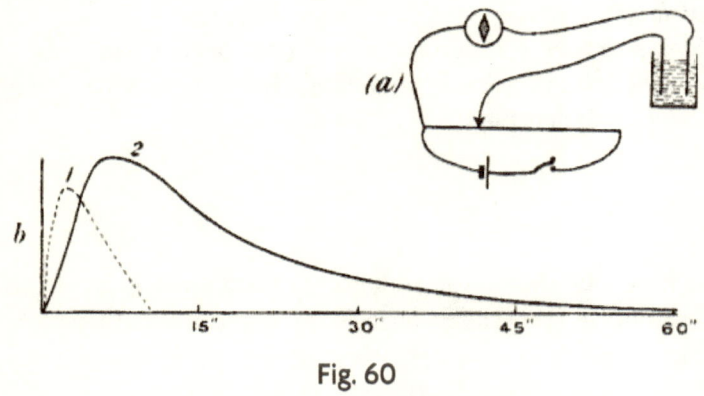

Fig. 60

(a) Arrangement for applying a short-lived E.M.F.
(b) Difference in the periods of recovery:
   (1) from instantaneous E.M.F.; and
   (2) that caused by mechanical stimulus.

The record was then taken as follows. The recording drum had a fast speed of six inches per minute, one of the small subdivisions representing a second. The battery contact in the main potentiometer circuit was made for a quarter of a second as just mentioned and a record taken of the effect of a short-lived E.M.F. on the circuit containing the cell.

A record was next taken of the E.M. variation produced in the cell by a single stimulus. It will be seen on comparison of the two records that the maximum effect took place relatively later in the case of mechanical stimulus, and that whereas the galvanometer recovery in the former case took place in

11 seconds, the recovery in the latter was not complete till after 60 seconds (*Figure 60b*). This shows that it takes some time for the effect of stimulus to attain its maximum, and that the effect does not disappear till after the lapse of a certain interval. The time of recovery from strain depends on the intensity of stimulus. It takes a longer time to recover from a stronger stimulus. But, other things being equal, successive recovery periods from successive stimulations of equal intensity are, generally speaking, the same.

We may now study the influence of any change in external conditions by observing the modifications it produces in the normal curve.

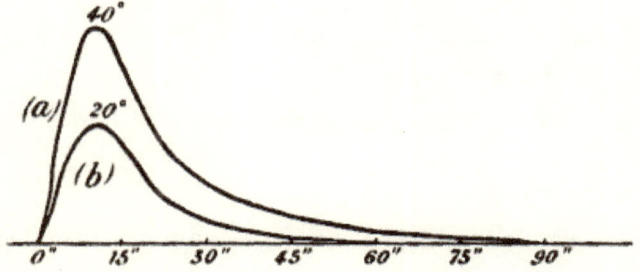

Fig. 61 **Prolongation of period of recovery after overstrain**

Recovery is complete in 60 seconds when the stimulus is due to 20° vibration. But with stronger stimulus of 40° vibration, the period of recovery is prolonged to 90 seconds.

## Prolongation of period of recovery by overstrain

The pair of records given in *Figure 61* shows how recovery is delayed, as the effect of overstrain. Curve (*a*) is for a single stimulus due to a vibration of amplitude 20°, and curve (*b*) for a stimulus of 40° amplitude of vibration. It will be noticed how relatively prolonged is the recovery in the latter case.

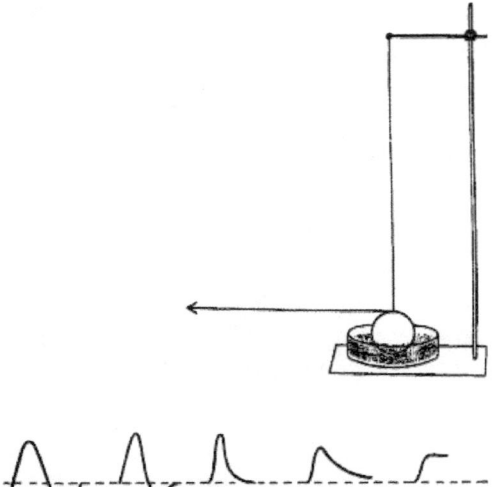

Fig. 62 **Model showing the effect of friction**

## Molecular model

We have seen that the electric response is an outward expression of the molecular disturbance produced by the action of the stimulus. The rising part of the response curve thus exhibits the effect of molecular upset, and the falling part, or recovery, the restoration to equilibrium. The mechanical model (*Figure 62*) will help us to visualise many complex response phenomena. The molecular model consists of a torsional pendulum – a wire with a dependent sphere. By the stimulus of a blow there is produced a torsional vibration – a response followed by recovery. The writing lever attached to the pendulum records the response curves. The form of these curves, stimulus remaining constant, will be modified by friction; the less the friction, the greater is the mobility. The friction may be varied by more or less raising or lowering a vessel of sand touching the pendulum. By varying the friction the following curves were obtained.

(*a*) When there is little friction we get an after-oscillation, to which we have the corresponding phenomenon in the retinal after-oscillation (compare *Figure 105*).

(*b* and *c*) If the friction is increased, there is a damping of oscillation. In (*c*) we get recovery curves similar to those found in nerve, muscle, plant, and metal.

(*d*) If the friction is still further increased the maximum is reached much later, as will be seen in the increasing slant of the rising part of the curve; the height of response is diminished and the period of recovery very much prolonged by partial molecular arrest. The curve (*d*) is very similar to the 'molecular arrest' curve obtained by a small dose of chemical reagents which act as 'poison' on living tissue or on metals (compare *Figure 93a*).

(*e*) When the molecular mobility is further decreased there is no recovery (compare *Figure 93b*).

Still further increase of friction completely arrests the molecular pendulum, and there is no response.

From what has been said, it will be seen that if in any way the friction is diminished or mobility increased the response will be enhanced. This is well exemplified in the heightened response after annealing (*Figure 58*) and after preliminary vibration (*Figures 81, 82*).

Possibly connected with this may be the increased responses exhibited by the action of stimulants (*Figures 89, 90*).

## Reduction of molecular sluggishness attended:
### (1) by quickened recovery

Sometimes, after a cell has been resting for too long a period, especially on cold days, the wire gets into a sluggish condition, and the period of recovery is thereby prolonged. But successive vibrations gradually remove this inertness, and recovery is then hastened. This is shown in the accompanying curves, *Figure 63*, where (*a*) exhibits only very partial recovery even after the expiration of 60 seconds, whereas when a few vibrations had been given recovery was entirely completed in 47 seconds (*b*). There was here little change in the height of response.

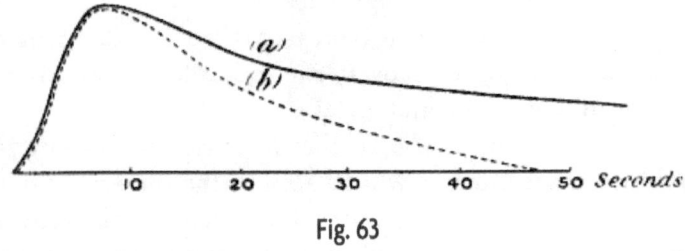

Fig. 63

(*a*) Slow recovery of a wire in a sluggish condition.
(*b*) Quickened recovery in the same wire after a few vibrations.

## Or (2) by heightened response

The removal of sluggishness by vibration, resulting in increased molecular mobility, is in other instances accompanied by increase in the height of response, as will be seen from the two sets of records which follow (*Figure 64*). Cold, due to prevailing frosty weather, had made the wires in the cell somewhat lethargic.

The records in (*a*) were the first taken on the day of the experiment. The amplitudes of vibration were 45°, 90°, and 135°. In (*b*) are given the records of the next series, which are in every case greater than those of (*a*).

This shows that previous vibration, by conferring increased mobility, had heightened the response. In this case, removal of molecular sluggishness is accompanied by greater intensity of response, without much change in the period of recovery. In connection with this it must be remembered that greater strain consequent on heightened response has a general tendency to a prolongation of the period of recovery.

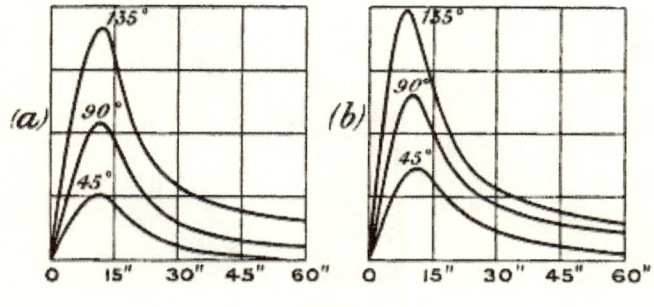

Fig. 64

*(a)* Three sets of responses for 45°, 90°, and 135° vibration in a sluggish wire.

*(b)* The next three sets of responses in the same wire; increased mobility conferred by previous vibration has heightened the response.

It is thus seen that when the wire is in a sluggish condition, successive vibrations confer increased molecular mobility, which finds expression in quickened recovery or heightened response.

## Effect of temperature

Similar considerations lead us to expect that a moderate rise of temperature will be conducive to increase of response. This is exhibited in the next series of records.

The wire at the low temperature of 5°C happened to be in a sluggish condition, and the responses to vibrations of 45° to 90° in amplitude were feeble. Tepid water at 30°C was now substituted for the cold water in the cell, and the responses underwent a remarkable enhancement. But the excessive molecular disturbance caused by the high temperature of 90°C produced a great diminution of response (*Figure 65*).

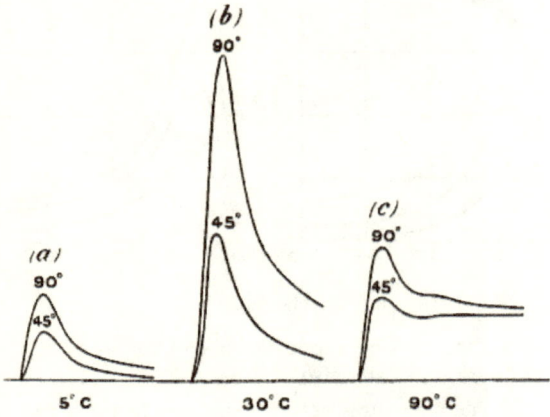

Fig. 65 **Responses of a wire to amplitudes of vibration 45° and 90°**

*(a)* Responses when the wire was in a sluggish condition at temperature of 5°C.

*(b)* Enhanced response at 30°C.

*(c)* Diminution of response at 90°C.

## Diphasic variation

It has already been said that if two points A and B are in the same physico-chemical condition, then a given stimulus will give rise to similar excitatory electric effects at the two points. If the galvanometer deflection is 'up' when A alone is excited, the excitation of B will give rise to a downward deflection. When the two points are simultaneously excited the electric variation at the two points will *continuously* balance each other. Under such conditions there will be no resultant deflection. But if the intensity of stimulation of one point is relatively stronger, then the balance will be disturbed, and a resultant deflection produced whose sign and magnitude can be found independently by the algebraical summation of the individual effects of A and B.

It has also been shown that a balancing point for the block, which is approximately near the middle of the wire, may be found so that the vibrations of A and B through the same amplitude produce equal and opposite deflection. Simultaneous vibration of both will give no resultant current; when the block is abolished and the wire is vibrated as a whole, there will still be no resultant current, inasmuch as similar excitations are produced at A and B.

After obtaining the balance, if we apply an exciting reagent such as $Na_2CO_3$ at one point, and a depressing reagent such as KBr at the other, the responses will now become unequal, the more excitable point giving a stronger deflection.

We can, however, make the two deflections equal by increasing the amplitude of vibration of the less sensitive point. The two deflections may thus be rendered equal and opposite, but the time relations – the latent period, the time rate for attaining the maximum excitation and recovery from that effect – will no longer be the same in the two cases. There would therefore be no continuous balance, and we obtain instead a very interesting diphasic record. I give below an exact reproduction of the response curves of A and B recorded on a fast-moving drum.

It will be remembered that one point was touched with $Na_2CO_3$ and the other with KBr. By suitably increasing the amplitude of vibration of the less sensitive, the two deflections were rendered approximately equal. The records of A and B were at first taken separately (*Figure 66a*). It will be noticed that the maximum deflection of A was attained earlier than that of B. The resultant curve R' was obtained by summation.

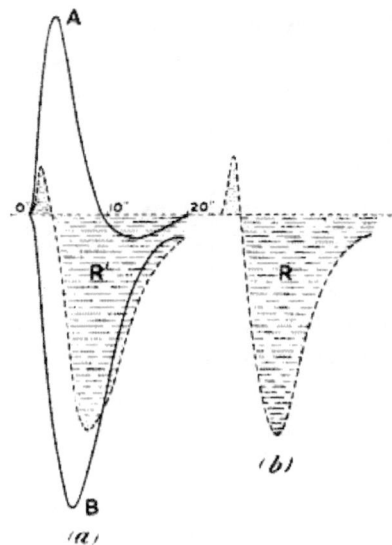

Fig. 66 **Diphasic variation**

*(a)* Records of A and B obtained separately.
R' is the resultant by algebraical summation.

*(b)* Diphasic record obtained by simultaneous stimulation of A and B.

After taking the records of A and B separately, a record of the resultant effect R due to simultaneous vibration of A and B was next taken. It gave the curious two-phased response – a positive effect followed by negative after-vibration, practically similar to the resultant curve R' (*Figure 66b*).

The positive portion of the curve is due to A effect and the negative to B. If by any means, say by either increasing the amplitude of vibration of A or increasing its sensitiveness, the response of A is very greatly enhanced, then the positive effect would be predominant and the negative effect would become inconspicuous. When the two constituent responses are of the same order of magnitude, we shall have a positive response followed by a negative after-vibration; the first twitch will belong to the one which responds earlier. If the response of A is very much reduced, then the positive effect will be reduced to a mere twitch and the negative effect will become predominant.

I give a series of records, *Figure 67*, in which these three principal types are well exhibited, the two contacts having been rendered unequally excitable by solutions of the two reagents KBr and $Na_2CO_3$. A and B were vibrated simultaneously and records taken.

(*a*) First, the relative response of B (downward) is increased by increasing its amplitude of vibration. The amplitude of vibration of A was throughout maintained constant. The negative or downward response is now very conspicuous, there being only a mere preliminary indication of the positive effect.

(*b*) The amplitude of vibration of B is now slightly reduced, and we obtain the diphasic effect.

(*c*) The intensity of vibration of B is diminished still further, and the negative effect is seen reduced to a slight downward after-vibration, the positive up-curve being now very prominent (*Figure 67*).

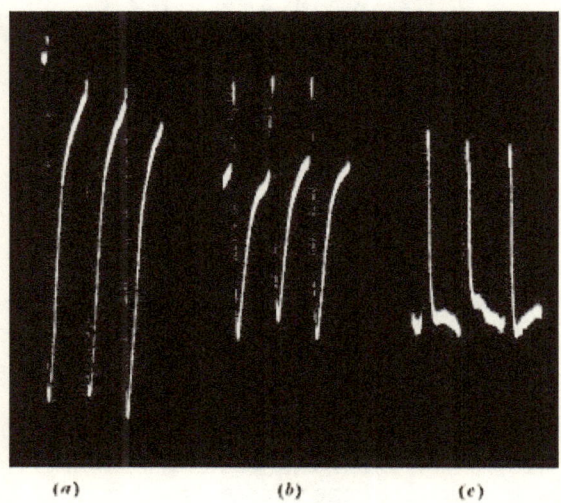

Fig. 67 **Negative, diphasic, and positive resultant response**

## Continuous transformation from negative to positive

I have shown the three phases of transformation, the intensity of one of the constituent responses being varied by altering the intensity of disturbance.

In the following record (*Figure 68*) I succeeded in obtaining a continuous transformation from positive to negative phase by a continuous change in the relative sensitiveness of the two contacts.

I found that traces of after-effect due to the application of $Na_2CO_3$ remain for a time. If the reagent is previously applied to an area and the traces of the carbonate then washed off, the increased sensitiveness conferred disappears gradually. Again, if we apply $Na_2CO_3$ solution to a fresh point, the sensitiveness gradually increases. There is another further interesting point to be noticed: the beginning of response is earlier when the application of $Na_2CO_3$ is fresh.

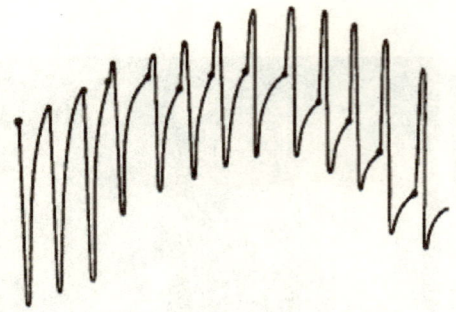

**Fig. 68 Continuous transformation from negative to positive through intermediate diphasic response**

Thick dots represent the times of application of successive stimuli.

We have thus a wire held at one end, and successive uniform vibrations at intervals of one minute imparted to the wire as a whole, by means of a vibration head on the other end.

Owing to the after-effect of previous application of

$Na_2CO_3$ the sensitivity of B is at the beginning great, hence the three resultant responses at the beginning are negative or downward.

Dilute solution of $Na_2CO_3$ is next applied to A. The response of A (up) begins earlier and continues to grow stronger and stronger. Hence, after this application, the response shows a preliminary positive twitch of A followed by negative deflection of B. The positive grows continuously. At the fifth response the two phases, positive and negative, become equal, after that the positive becomes very prominent, the negative being reduced as a feeble after-vibration.

It need only be added here that the diphasic variations as exhibited by metals are in every way counterparts of similar phenomena observed in animal tissues.

# Chapter 14
# INORGANIC RESPONSE – FATIGUE, STAIRCASE, AND MODIFIED RESPONSE

*Fatigue in metals – Fatigue under continuous stimulation – Staircase effect – Reversed responses due to molecular modification in nerve and metal, and their transformation into normal after continuous stimulation – Increased response after continuous stimulation.*

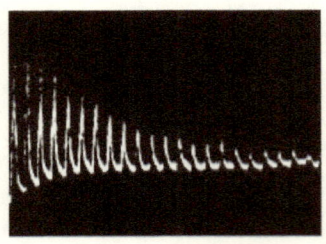

Fig. 69 **Fatigue in muscle (Waller)**

## Fatigue

In some metals, as in muscle and in plant, we find instances of that progressive diminution of response which is known as fatigue (*Figure 69*). The accompanying record shows this in platinum (*Figure 70*). It has been said that tin is practically indefatigable. We must, however, remember that this is a question of degree only. Nothing is *absolutely* indefatigable. The exhibition of fatigue depends on various conditions. Even in tin, then, I obtained the characteristic fatigue curve with a specimen which had been in continuous use for many days (*Figure 71*). While discussing the subject of fatigue in plants,

I have adduced considerations which showed that the residual effect of strain was one of the main causes for the production of fatigue. This conclusion receives independent support from the records obtained with metals.

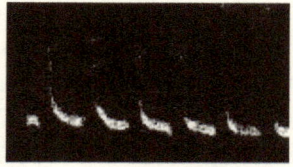

Fig. 70 **Fatigue in platinum**

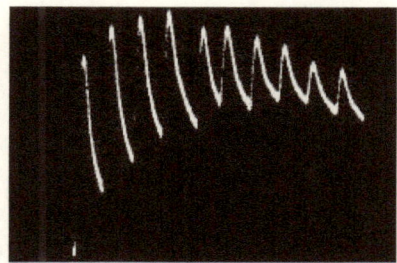

Fig. 71 **Fatigue shown by tin wire which had been continuously stimulated for several days**

In this connection the important fact is that the various typical fatigue effects exhibited in living substances are exactly reproduced in metals, where there can be question neither of fatigue-product producing fatigue effects, nor of those constructive processes by which they might be removed. We have seen, both in muscles and in plants, that if sufficient time for complete recovery is allowed between each pair of stimuli, the heights of successive responses are the same, and there is no apparent fatigue (see p.67). But the height of response diminishes as the excitation interval is shortened. We find the same thing in metals. Below is given a record taken with tin (*Figure 72*). Throughout the experiment the amplitude of vibration was maintained constant, but in (*a*) the interval between consecutive stimuli was one minute, while in (*b*) this was reduced to 30 seconds. A diminution of height immediately occurs. On restoring the original rhythm as in (*c*), the responses revert to their first large value.

Thus we see that when the wire has not completely recovered, its responses, owing to residual strain, undergo diminution. Height of response is thus decreased by incomplete recovery. If then sufficient time is not allowed for perfect recovery, we can understand how, under certain circumstances, the residual strain would progressively increase with repetition of stimulus, and thus there would be a progressive diminution of height of response or fatigue.

Again, we saw in the last chapter that increase of strain necessitates a longer period of recovery. Thus the longer a wire is stimulated, the more and more overstrained it becomes, and it therefore requires a gradual prolongation of the interval between the successive stimuli, if recovery is to be complete. This interval, however, being maintained constant, the recovery periods virtually undergo a gradual reduction, and successive recoveries become more and more incomplete. These considerations may be found to afford an insight into the progressive diminution of response in fatigued substances.

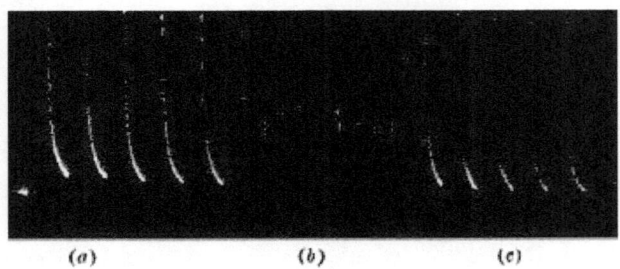

### Fig. 72 **Diminution of response due to shortening the period of recovery**

The stimulus is maintained constant. In *(a)* the interval between two successive stimuli is one minute, in *(b)* it is half a minute, and in *(c)* it is again one minute. The response in *(b)* is feebler than in either *(a)* or *(c)*.

## Fatigue under continuous stimulation

Fatigue is perhaps best shown under continuous stimulation. For example, in muscles, when fresh and not fatigued, the top of the tetanic curve is horizontal, or may even be ascending, but with long-continued stimulation the curve declines. The rapidity of this decline depends on the nature of the muscle and its previous condition.

In metals I have found exactly parallel instances. In tin, so little liable to fatigue, the top of the curve is horizontal or ascending; or it may exhibit a slight decline. But the record with platinum shows the rapid decline due to fatigue (*Figure 73*).

Fig. 73

*(a)* The top of response curve under continuous stimulation in tin is horizontal or ascending.
*(b)* In platinum there is rapid decline owing to fatigue.

Taking any of these instances, say that in which fatigue is most prominent, it is found that a short period of rest restores the original intensity of response. This affords additional proof of the fact that fatigue is due to overstrain, and that this strain, with its sign of attendant fatigue, disappears with time.

## Staircase effect

We shall now discuss an effect which appears to be the direct opposite of fatigue. This is the curious phenomenon known to physiologists as the 'staircase' effect, in which successive uniform stimuli produce a series of increasing responses. This is seen under particular conditions in the response of certain muscles (*Figure 74a*). It is also observed sometimes even in nerve, which otherwise, generally speaking, gives uniform responses. Of this effect, no satisfactory theory has as yet been offered. It is in direct contradiction to that theory which supposes that each stimulus is followed by dissimilation or breakdown of the tissue, reducing its function below par. For in these cases the supposed dissimilation is followed not by a decrease but by an increase of functional activity.

This 'staircase effect' I have shown to be occasionally exhibited by plants. I have also found it in metals. In the last chapter we have seen that a wire often falls, especially after resting for a long time, into a state of comparative sluggishness, and that this molecular inertness then gradually gives way to increased mobility under stimulation.

As a consequence, an increased response is thus obtained. I give in *Figure 74b*, a series of responses to uniform stimuli, exhibited by platinum which had been at rest for some time. This effect is very clearly shown here. So we see that in a substance which has previously been in a sluggish condition, stimulation confers increased mobility. Response thus reaches a maximum, but continued stimulation may afterwards produce overstrain, and the subsequent responses may then show a decline. This consideration will explain certain types of responses exhibited by muscles, where the first part of the series exhibits a staircase increase followed by declining responses of fatigue.

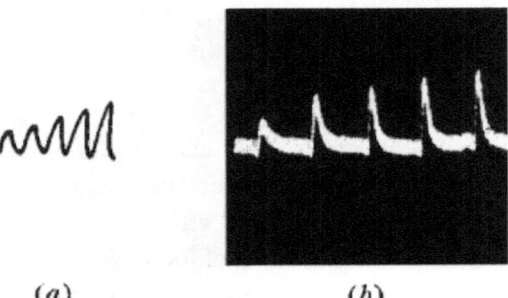

(a) (b)

Fig. 74 **The 'staircase' effect**

*(a)* in muscle (after Engelmann). *(b)* in metal.

## Reversed response due to molecular modification and its transformation into normal after continuous stimulation:

### (1) in nerve

Reference has already been made to the fact that a nerve which, when fresh, exhibited the normal negative response, will often, if kept for some time in preservative saline, undergo a molecular modification, after which it gives a positive variation. Thus while the response given by fresh nerve is *normal* or negative, a stale nerve gives *modified*, i.e. reversed or positive, response. This peculiar modification does not always occur, yet is too frequent to be considered abnormal. Again, when such a nerve is subjected to tetanisation or continuous stimulation, this modified response tends once more to become normal.

It is found that not only tetanisation, but also $CO_2$ has the power of converting the modified response into normal. Hence it has been suggested that the conversion under tetanisation of modified response to normal, in stale nerve, is due to a hypothetical evolution of $CO_2$ in the nerve during stimulation.[16]

### (2) In metals

I have, however, met with exactly parallel phenomena in metals, where, owing to some molecular modification, the responses became reversed, and where, under continuous stimulation, though here there could be no possibility of the evolution of $CO_2$, they tended again to become normal.

If after mounting a wire in a cell filled with water, it is set aside for too long a time, I have sometimes noticed that it undergoes a certain modification, owing to which its response ceases to be normal and becomes reversed in sign. I have obtained this effect with various metals, for instance lead and tin, and even with the chemically inactive substance platinum.

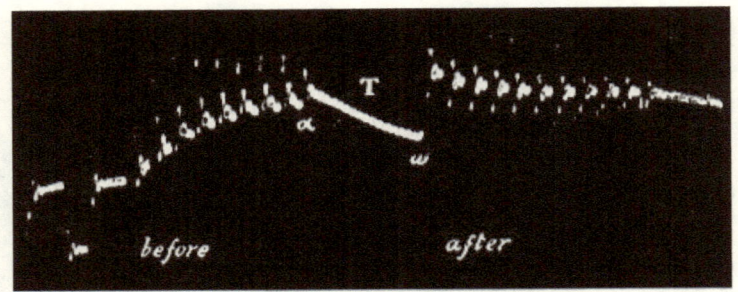

Fig. 75 **Abnormal positive (up) response in nerve converted into normal (down) response after continuous stimulation 'T'** (Waller)

The galvanometer is not dead-beat, and shows after-oscillation.

The subject will be made clearer if we first follow in detail the phenomenon exhibited by modified nerve giving this abnormal response. The normal responses in nerve are usually represented by 'down' and the reversed abnormal responses by 'up' curves. In the modified nerve, then, the abnormal responses are 'up' instead of the normal 'down.' The record of such abnormal response in the modified nerve is shown in *Figure 75*. It will be noticed that in this, the successive responses are undergoing a diminution, or tending towards the normal. After continuous stimulation or tetanisation (T), it will be seen that the abnormal or 'up' responses are converted into normal or 'down.'

I shall now give a record which will exhibit an exactly similar transformation from the abnormal to normal response after continuous stimulation. Here the normal responses are represented by 'up' and the abnormal by 'down' curves. This record was given by a tin wire, which had been molecularly modified (*Figure 76*). We have at first the abnormal responses; successive responses are undergoing a diminution or tending towards the normal; after continuous stimulation (T), the subsequent responses are seen to have become normal. Another record, obtained with platinum, shows the same phenomenon (*Figure 77*).

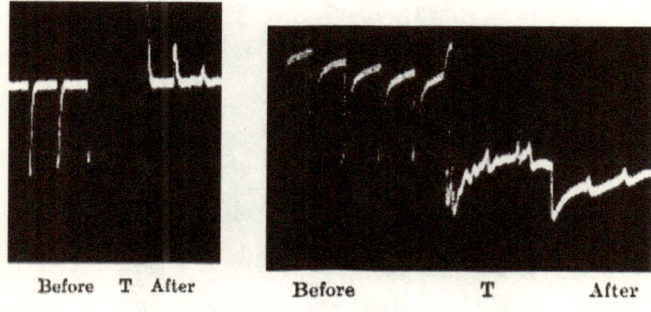

Before   T   After            Before              T              After

**Figs 76 and 77**

Abnormal 'down' response in tin (*fig. 76*) and in platinum (*fig. 77*) transformed into normal 'up' response, after continuous stimulation, T.

On placing the three sets of records – nerve, tin, and platinum – side by side, it will be seen how essentially similar they are in every respect.[17]

This reversion to normal is seen to have appeared in a pronounced manner after rapidly continuous stimulation, in the process of which the modified molecular condition must in some way have reverted to the normal.

Fig. 78 **The gradual transition from abnormal to normal response in platinum**

The transition will be seen to have commenced at the third and ended at the seventh, counting from the left.

Being desirous to trace this change gradually taking place, I took a platinum wire cell giving modified responses, and obtained a series of records of effects of individual stimuli continued for a long time. In this series, the points of transition from modified response to normal are clearly seen (*Figure 78*).

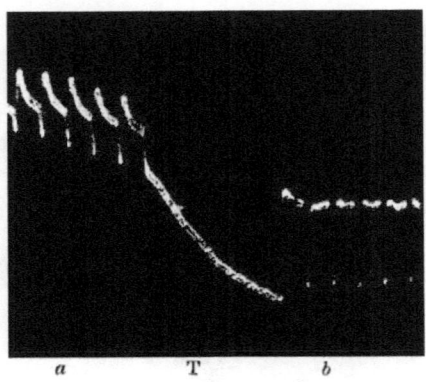

Fig. 79 **The normal response 'a' in nerve enhanced to 'b' after continuous stimulation 'T'** (Waller)

The normal response in nerve is recorded 'down.'

## Increased response after continuous stimulation

We have seen that responses to uniform stimuli sometimes show a staircase increase, apparently owing to the gradual removal of molecular sluggishness. Possibly analogous to this is the increase of response in nerve after continuous stimulation or tetanisation, observed by Waller (*Figure 79*). Like the staircase effect, this contravenes the commonly accepted theory of the dissimilation of tissue by stimulus, and the consequent depression of response. It is suggested by Waller that this increase of response after tetanisation may be due to the hypothetical evolution of $CO_2$ to which allusion has previously been made.

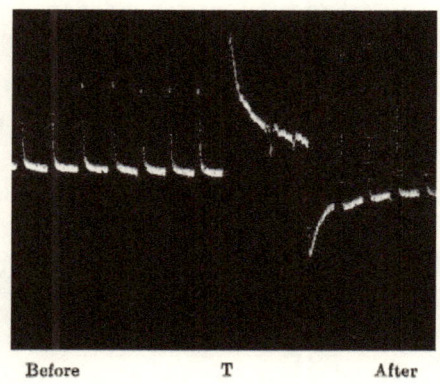

Fig. 80 **Enhanced response in platinum after continuous stimulation 'T'**

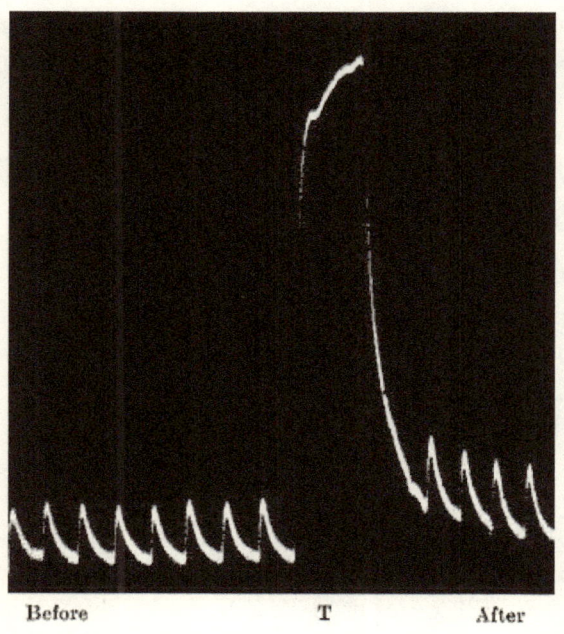

Fig. 81 **Enhanced response in tin after continuous stimulation 'T'**

But there is an exact correspondence between this phenomenon and that exhibited by metals under similar conditions. I give here two sets of records (*Figures 80, 81*), one obtained with platinum and the other with tin, which demonstrate how the response is enhanced after continuous stimulation in a manner exactly similar to that noticed in the case of nerve.

The explanation which has been suggested with regard to the staircase effect – increased molecular mobility due to removal of sluggishness by repeated stimulation – would appear to be applicable in this case also. It would appear, then, that in all the phenomena which we have studied under the heading of the 'staircase effect', increase of response after continuous stimulation, and fatigue, there is a similarity between the observations made upon the response of muscle and nerve on the one hand, and that of metals on the other. Even in their abnormalities we have seen an agreement.

But amongst these phenomena themselves, though at first sight so diverse, there is some kind of continuity. Calling *all* normal response *positive*, for the sake of convenience, we observe its gradual modification, corresponding to changes in the molecular condition of the substance.

Beginning with that case in which molecular modification is extreme, we find a maximum variation of response from the normal, that is to say, to *negative*.

Continued stimulation, however, brings back the molecular condition to normal, as evidenced by the progressive lessening of the negative response, culminating in reversion to the normal *positive*. This is equally true of nerve and metal.

In the next class of phenomena, the modification of molecular condition is not so great. It now exhibits itself merely as a relative inertness, and the responses, though positive, are feeble. Under continued stimulation, they increase in the same direction as in the last case, that is to say, from less positive to *more positive*, being the reverse of fatigue. This is evidenced alike by the staircase effect and by the increase of response after tetanisation, seen not only in nerve but also in platinum and tin.

The substance may next be in what we call the normal condition. Successive uniform stimuli now evoke uniform and equal positive responses, that is to say, there is no fatigue. But after intense or long-continued stimulation, the substance is overstrained. The responses now undergo a change from positive to *less positive*; fatigue, that is to say, appears.

Again, under very much prolonged stimulation the response may decline to zero, or even undergo a reversal to *negative*, a phenomenon which we shall find instanced in the reversed response of retina under the long-continued stimulus of light.

We must then recognise that a substance may exist in various molecular conditions, whether due to internal changes or to the action of stimulus. The responses give us indications of these conditions. A complete cycle of molecular modifications can be traced, from the abnormal negative to the normal positive, and then again to negative seen in reversal under continuous stimulation.

## Footnotes

16. "Considering that we have no previous evidence of any chemical or physical change in tetanised nerve, it seems to me not worthwhile pausing to deal with the criticism that it is not $CO_2$, but 'something else' that has given the result." – Waller, *Animal Electricity*, p.59. That this phenomenon is nevertheless capable of physical explanation will be shown presently.

17. In order to explain the phenomena of electric response, some physiologists assume that the negative response is due to a process of dissimilation, or breakdown, and the positive to a process of assimilation, or building up, of the tissue. The modified or positive response in nerve is thus held to be due to assimilation; after continuous stimulation, this process is supposed to be transformed into one of dissimilation, with the attendant negative response.

    How arbitrary and unnecessary such assumptions are will become evident, when the abnormal and normal responses, and their transformation from one to the other, are found repeated in all details in metals, where there can be no question of the processes of assimilation or dissimilation.

# Chapter 15
# INORGANIC RESPONSE – RELATION BETWEEN STIMULUS AND RESPONSE – SUPERPOSITION OF STIMULI

*Relation between stimulus and response – Magnetic analogue – Increase of response with increasing Stimulus – Threshold of response – Superposition of Stimuli – Hysteresis.*

## Relation between stimulus and response

We have seen what extremely uniform responses are given by tin, when the intensity of stimulus is maintained constant. Hence it is obvious that these phenomena are not accidental, but governed by definite laws. This fact becomes still more evident when we discover how invariably response is increased by increasing the intensity of stimulus.

Electrical response is due, as we have seen, to a molecular disturbance, the stimulus causing a distortion from a position of equilibrium. In dealing with the subject of the relation between the disturbing force and the molecular effect it produces, it may be instructive to consider certain analogous physical phenomena in which molecular deflections are also produced by a distorting force.

## Magnetic analogue

Let us consider the effect that a magnetising force produces on a bar of soft iron. It is known that each molecule in such a bar is an individual magnet. The bar as a whole, nevertheless, exhibits no external magnetisation. This is held to be due to the fact that the molecular magnets are turned either in haphazard directions or in closed chains, and there is therefore no resultant polarity. But when the bar is subjected to a magnetising force by means, say, of a solenoid carrying electrical current, the individual molecules are elastically deflected, so that all the molecular magnets tend to place themselves along the lines of magnetising force. All the north poles thus point more or less one way, and the south poles the other. The stronger the magnetising force, the nearer do the molecules approach a perfect alignment, and the greater is the induced magnetisation of the bar.

The intensity of this induced magnetisation may be measured by noting the deflection it produces on a freely suspended magnet in a magnetometer.

The force which produces that molecular deflection, to which the magnetisation of the bar is immediately due, is the magnetising current flowing round the solenoid. The magnetisation, or the molecular effect, is measured by the deflection of the magnetometer. We may express the relation between cause and effect by a curve in which the abscissa ($x$) represents the magnetising current, and the ordinate ($y$) the magnetisation produced (*Figure 82*).

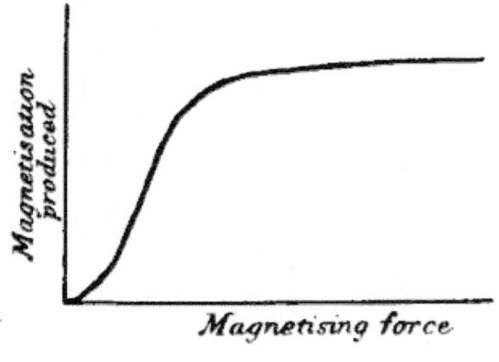

Fig. 82 **Curve of magnetisation**

In such a curve we may roughly distinguish three parts. In the first, where the force is feeble, the molecular deflection is slight. In the next, the curve is rapidly ascending, i.e. a small variation of impressed force produces a relatively large molecular effect. And lastly, a limit is reached, as seen in the third part, where increasing force produces very little further effect. In this cause-and-effect curve, the first part is slightly convex to the abscissa, the second straight and ascending, and the third concave.

## Increase of response with increasing stimulus

We shall find in dealing with the relation between the stimulus and the molecular effect – i.e. the response – something very similar.

On gradually increasing the intensity of stimulus, which may be done, as already stated, by increasing the amplitude of vibration, it will be found that, beginning with feeble stimulation, this increase is at first slight, then more pronounced, and lastly shows a tendency to approach a limit. In all this we have a perfect parallel to corresponding phenomena in animal and vegetable response. We saw that the proper investigation of this subject was much complicated, in the case of animal and vegetable tissues, by the appearance of fatigue.

The comparatively indefatigable nature of tin causes it to offer great advantages in the pursuit of this inquiry. I give below two series of records made with tin. The first record, *Figure 83*, is for increasing amplitudes from 5° to 40° by steps of 5°. The stimuli are imparted at intervals of one minute. It will be noticed that whereas the recovery is complete in one minute when the stimulus is moderate, it is not quite complete when the stimulus is stronger. The recovery from the effect of stronger stimulus is more prolonged. Owing to want of complete recovery, the base line is tilted slightly upward. This slight displacement of the zero line does not materially affect the result, provided the shifting is slight.

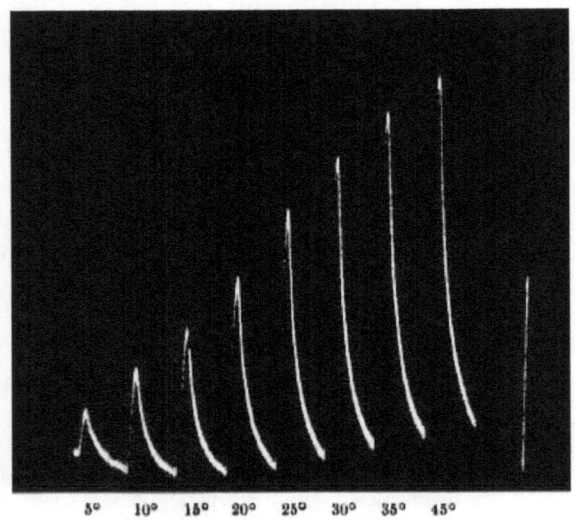

Fig. 83 **Records of responses in tin with increasing stimuli, amplitudes of vibration from 5° to 40°**

The vertical line to the right represents .1 volt.

## Table showing the increasing electric response due to increasing amplitude of vibration

| Vibration amplitude | e.m. variation |
|---|---|
| 5° | .024 volt |
| 10° | .057 volt |
| 20° | .111 volt |
| 25° | .143 volt |
| 30° | .170 volt |
| 35° | .187 volt |
| 40° | .204 volt |

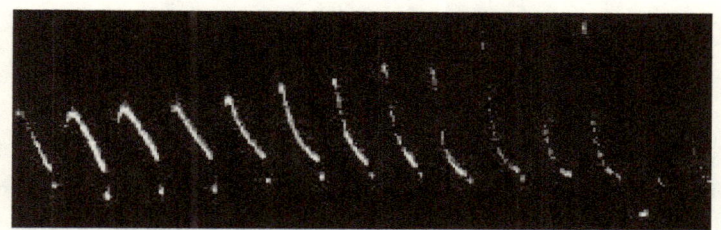

Fig. 84 **A second set of records with a different specimen of tin**

The amplitudes of vibration are increased by steps of 10°, from 20° to 160°. (The deflections are reduced by interposing a high external resistance.)

The next figure (*Figure 84*) gives record of responses through a wider range. For accurate quantitative measurements it is preferable to wait till the recovery is complete. We may accomplish this within the limited space of the recording photographic plate by making the record for one minute. During the rest of recovery, the clockwork moving the plate is stopped and the galvanometer spot of light is cut off. Thus the next record starts from a point of completed recovery, which will be noticed as a bright spot at the beginning of each curve. With stimulation of high intensity, a tendency will be noticed for the responses to approach a limit.

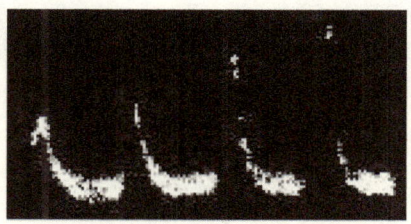

Fig. 85 **Effect of superposition on tin**

A single stimulus produces the feeble effect shown in the first response. Superposition of 5, 9, 13 such stimuli produce the succeeding stronger responses.

## Threshold of response

There is a minimum intensity of stimulus below which there is hardly any visible response. We may regard this point as the threshold of response. Though apparently ineffective, the subliminal stimuli produce some latent effect, which may be demonstrated by their additive action. The record in *Figure 85* shows how individually feeble stimuli become markedly effective by superposition.

## Superposition of stimuli

The cumulative effect of succeeding stimuli can be seen in the above. The fusion of effect will be incomplete if the frequency of stimulation be not sufficiently great; but it will tend to be more complete with higher frequency of stimulation (*Figure 86*). We have here a parallel case to the complete and incomplete tetanus of muscles, under similar conditions.

By the addition of these rapidly succeeding stimuli, a maximum effect is produced, and further stimulation adds nothing to this. The effect is balanced by a force of restitution. The response curve thus rises to its maximum, after which the deflection maintained, so long as the vibration is kept up.

It was found that increasing intensities of single stimuli produced correspondingly increased responses. The same is true also of groups of stimuli. The maximum effect produced by superposition of stimuli increases with the intensity of the constituent stimuli.

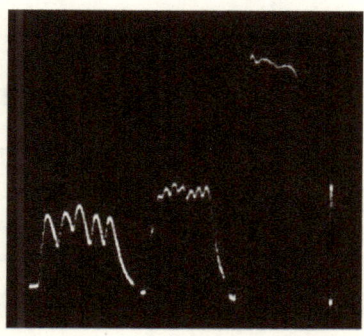

Fig. 86 **Incomplete and complete fusion of effect in tin**

As the frequency of stimulation is increased the fusion becomes more and more complete. Vertical line to the right represents .1 volt.

## Hysteresis

Allusion has already been made to the increased responsiveness conferred by preliminary stimulation (see p.152). Wanting to find out in what manner this is brought about, I took a series of observations for an entire cycle, that is to say, a series of observations were taken for maximum effects, starting from amplitude of vibration of 10° and ending in 100°, and backwards from 100° to 10°. Effect of hysteresis is very clearly seen (see A, *Figure 87*); there is a considerable divergence between the forward and return curves, the return curve being higher. On repeating the cycle several times, the divergence is very much reduced, and the wire on the whole is found to assume a more constant sensitiveness. In this steady condition, generally speaking, the sensitivity for smaller amplitude of vibration is found to be greater than at the very beginning, but the reverse is the case for stronger intensity of stimulation.

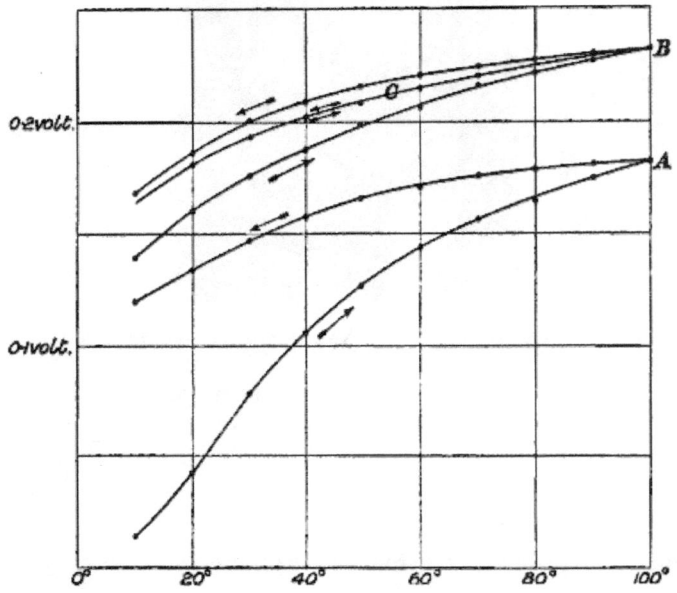

Fig. 87 **Cyclic curve for maximum effects showing hysteresis**

## Effect of annealing

I repeated the experiment with the same wire, after pouring hot water into the cell and allowing it to cool to the old temperature. From the cyclic curve (B, *Figure 87*) it will be seen:

(1) that the sensitiveness has become very much enhanced;

(2) that there is relatively less divergence between the forward and return curves. Even this divergence practically disappeared at the third cycle, when the forward and backward curves coincided (C, *Figure 87*).

The above results show in what manner the excitability of the wire is enhanced by purely physical means.

It is very curious to notice that addition of $Na_2CO_3$ solution produces an enhancement of responsive power similar to that produced by annealing; that is to say, not only is there a great increase of sensitivity, but there is also a reduction of hysteresis.

# Chapter 16
# INORGANIC RESPONSE - EFFECT OF CHEMICAL REAGENT

*Action of chemical reagents - Action of stimulants on metals - Action of depressants on metals - Effect of 'poisons' on metals - Opposite effect of large and small doses.*

We have seen that the ultimate evidence for the physiological character of electric response is its negation when the substance is subjected to those chemical reagents which act as poisons.

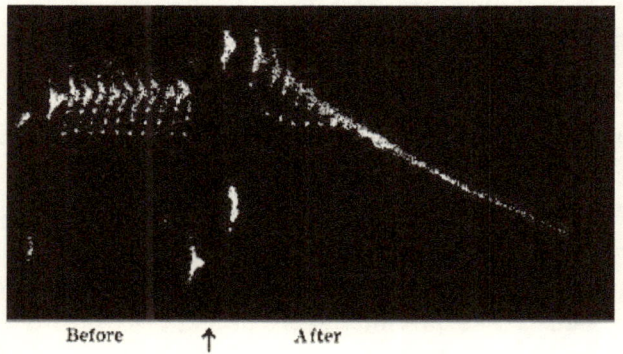

Before ↑ After

Fig. 88 **Action of poison in abolishing response in nerve** (Waller)

## Action of chemical reagents

Of these reagents, some are universal in their action, amongst which strong solutions of acids and alkalis, and salts such as mercuric chloride, may be cited. These act as powerful toxic agents, killing the living tissue, and causing electric response to disappear. (See *Figure 88*.)

It must, however, be remembered that there are again specific poisons which may affect one kind of tissue and not others. Poisons in general may be regarded as extreme examples of depressants. As an example of those which produce moderate physiological depression, potassium bromide may be mentioned, and this also diminishes electric response. There are other chemical reagents, on the other hand, which produce the opposite effect of increasing the excitability and causing a corresponding increase in electric response.

We shall now proceed to inquire whether the response of inorganic bodies is affected by chemical reagents, so that their excitability is increased by some, and depressed or abolished by others.

Should it prove to be so, the last test will have been fulfilled, and that parallelism which has been already demonstrated throughout a wide range of phenomena, between the electric response of animal tissues on the one hand, and that of plants and metals on the other, will be completely established.

## Action of stimulants on metals

We shall first study the stimulating action of various chemical reagents. The procedure is to take a series of normal responses to uniform stimuli, the electrolyte being water. The chemical reagent whose effect is to be observed is now added in small quantity to the water in the cell, and a second series of responses taken, using the same stimulus as before.

Generally speaking, the influence of the reagent is manifested in a short period, but there may be occasional instances where the effect takes some time to develop fully. We must remember that by the introduction of the chemical reagent some change may be produced in the internal resistance of the cell. The effect of this on the deflection is eliminated by interposing a very high external resistance (from one to five megohms) in comparison with which the internal resistance of the cell is negligible. The fact that the introduction of the reagent did not produce any variation in the total resistance of the circuit was demonstrated by taking two deflections, due to a definite fraction of a volt, before and after the introduction of the reagent. These deflections were found equal.

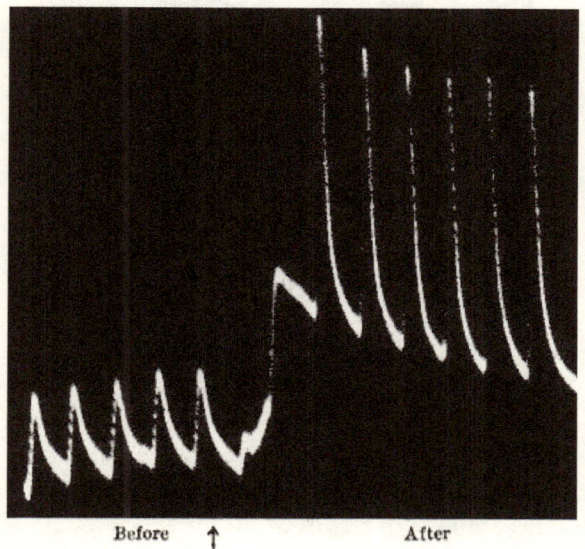

Fig. 89 **Stimulating action of $Na_2CO_3$ on tin**

I first give a record of the stimulating action of sodium carbonate on tin, which will become evident by a comparison of the responses before and after the introduction of $Na_2CO_3$ (*Figure 89*). The next record shows the effect of the same reagent on platinum (*Figure 90*).

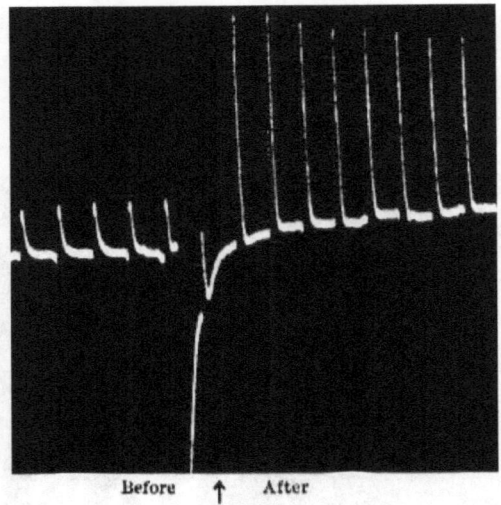

Fig. 90 **Stimulating action of Na$_2$CO$_3$ on platinum**

## Action of depressants

Certain other reagents, again, produce an opposite effect. That is to say, they *diminish* the intensity of response. The record given below (*Figure 91*) shows the depressing action of a 10% solution of KBr on tin.

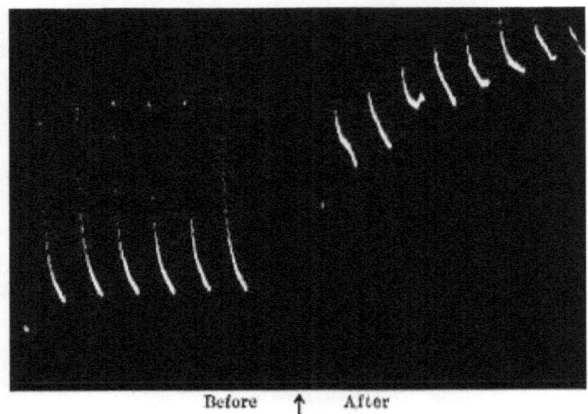

Fig. 91 **Depressing effect of KBr (10%) on the response of tin**

## Effect of 'poison'

Living tissues are killed, and their electric responses are at the same time abolished by the action of poisons. It is very curious that various chemical reagents are similarly effective in killing the response of metals.

I give below a record (*Figure 92*) to show how oxalic acid abolishes the response. The depressive effect of this reagent is so great that a strength of one part in 10,000 is often sufficient to produce complete abolition. Another notable point with reference to the action of this reagent is the persistence of after-effect. This will be clearly seen from an account of the following experiment.

The two wires A and B, in the cell filled with water, were found to give equal responses. The wires were now lifted off, and one wire B was touched with dilute oxalic acid. All traces of acid were next removed by rubbing the wire with cloth under a stream of water. On replacing the wire in the cell, A gave the usual response, whereas that of B was found to be abolished. The depression produced is so great and passes in so deep that I have often failed to revive the response, even after rubbing the wire with emery paper, by which the molecular layer on the surface must have been removed.

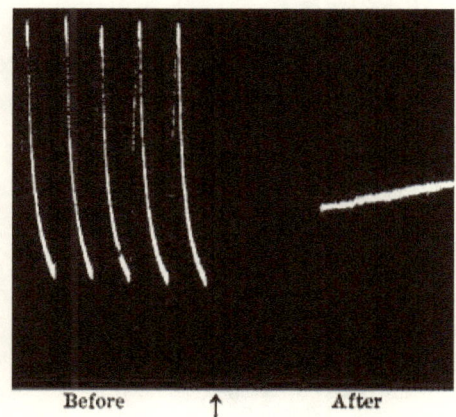

Fig. 92 **Abolition of response by oxalic acid**

We have seen in the molecular model (*Figure 62d,e*) how the attainment of maximum is delayed, the response diminished, and the recovery prolonged or arrested by increase of friction or reduction of molecular mobility.

It would appear that the reagents which act as poisons produced some kind of molecular arrest. The following records seen to lend support to this view.

If the oxalic acid is applied in large quantities, the abolition of response is complete. But on carefully adding just the proper amount I find that the first stimulus evokes a responsive electric twitch, which is less than the normal, and the period of recovery is very much prolonged from the normal one minute before, to five minutes after, the application of the reagent (*Figure 93a*).

In another record the arrest is more pronounced, i.e. there is now no recovery (*Figure 93b*). Note also that the maximum is attained much later.

Stimuli applied *after* the arrest produce no effect, as if the molecular mechanism became, as it were, clogged or locked up.

In connection with this it is interesting to note that the effect of veratrine poison on muscle is somewhat similar. This reagent not only diminishes the excitability, but causes a very great prolongation of the period of recovery.

In connection with the action of chemical reagents the following points are noteworthy.

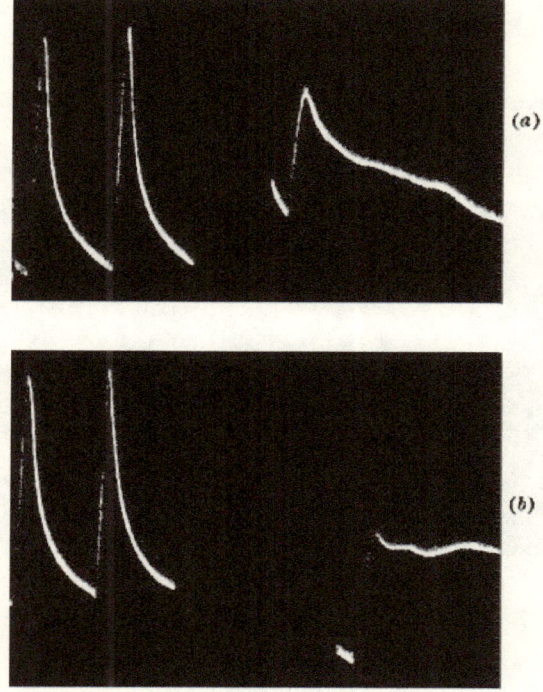

Fig. 93 **'Molecular arrest' by the action of 'poison'**

In each, curves to the left show the normal response, curve to the right shows the effect of poison. In (*a*) the arrest is evidenced by prolongation of period of recovery. In (*b*) there is no recovery.

(1) The effect of these reagents is not only to increase or diminish the height of the response curve, but also to modify the time relations. By the action of some the latent period is diminished, while others produce a prolongation of the period of recovery. Some curious effects produced by the change of time relations have been noticed in the account given of diphasic variation (see p.139).

(2) The effect produced by a chemical reagent depends to some extent on the previous condition of the wire.

(3) A certain time is required for the full development of the

effect. With some reagents the full effect takes place almost instantaneously, while with others the effect takes place slowly. Again the effect may with time reach a maximum, after which there may be a slight decline.

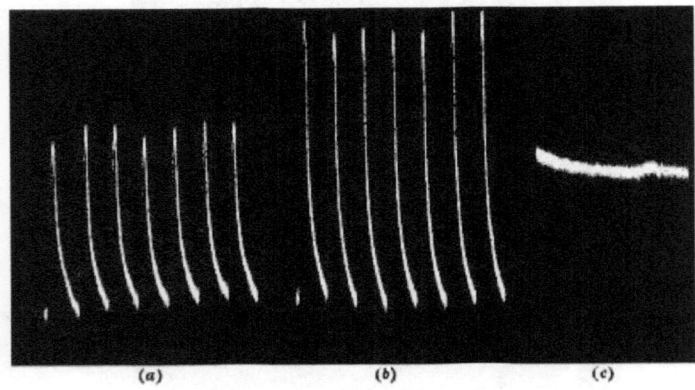

Fig. 94 **Opposite effects of small and large doses (tin)**

*(a)* is the normal response; *(b)* is the stimulating action of a small dose of potash (3 parts in 1,000); *(c)* is the abolition of response with a stronger dose (3 parts in 100).

(4) The after-effects of the reagents may be transitory or persistent; that is to say, in some cases the removal of the reagent causes the responses to revert to the normal, while in others the effect persists even after the removal of all traces of the reagent.

## Opposite effects of large and small doses

There remains a very curious phenomenon, known not only to students of physiological response but also known in medical practice, namely that of the opposite effects produced by the same reagent when given in large or in small doses. Here, too, we have the same phenomena reproduced in an extraordinary manner in inorganic response. The same reagent which becomes a 'poison' in large quantities may act as a stimulant when applied in small doses. This is seen in record *Figure 94*, in which (*a*) gives the normal responses in water; KHO solution was now added so as to make the strength three parts in 1,000, and (*b*) shows the consequent enhancement of response. A further quantity of KHO was added so as to increase the strength to three parts in 100. This caused a complete abolition of response (*c*).

It will thus be seen that as in the case of animal tissues and of plants, so also in metals, the electrical responses are exalted by the action of stimulants, lowered by depressants, and completely abolished by certain other reagents. The parallels are thus be to be found complete in every detail between the phenomena of response in the organic and the inorganic.

# Chapter 17
# ON THE STIMULUS OF LIGHT AND RETINAL CURRENTS

*Visual impulse: (1) chemical theory; (2) electrical theory – Retinal currents – Normal response positive – Inorganic response under stimulus of light – Typical experiment on the electrical effect induced by light.*

The effect of the stimulus of light on the retina is perceived in the brain as a visual sensation. The process by which the ether-wave disturbance causes this visual impulse is still very obscure. Two theories may be advanced in explanation.

## (1) Chemical theory

According to the first, or chemical, theory, it is supposed that certain visual substances in the retina are affected by light, and that vision originates from the metabolic changes produced in these visual substances. It is also supposed that the metabolic changes consist of two phases; the upward, constructive, or anabolic phase, and the downward, destructive, or katabolic phase.

Various visual substances by their anabolic or katabolic changes are supposed to produce the variations of sensation of light and colour. This theory, as will be seen, is very complex, and there are certain obstacles in the way of its acceptance.

It is, for instance, difficult to see how this very quick visual process could be due to a comparatively slow chemical action, consisting of the destructive breaking-down of the tissue, followed by its renovation. Some support was at first given to this chemical theory by the bleaching action of light on the visual purple present in the retina, but it has been found that the presence or absence of visual purple could not be essential to vision, and that its function, when present, is of only secondary importance. For it is well known that in the most sensitive portion of the human retina, the *fovea centralis*, the visual purple is wanting; it is also found to be completely absent from the retinæ of many animals possessing keen sight.

## (2) Electrical theory

The second, or electrical, theory supposes that the visual impulse is the concomitant of an electrical impulse; that an electrical current is generated in the retina under the influence of light, and that this is transmitted to the brain by the optic nerve. There is much to be said in favour of this view, for it is an undoubted fact that light gives rise to retinal currents, and that, conversely, an electrical current suitably applied causes the sensation of light.

### Retinal currents

Holmgren, Dewar, McKendrick, Kuhne, Steiner, and others have shown that illumination produces electric variation in a freshly excised eye. About this general fact of the electrical response there is a widespread agreement, but there is some difference of opinion as regards the sign of this response immediately on the application, cessation, and during the continuance of light. These slight discrepancies may be partly due to the unsatisfactory nomenclature – as regards use of terms *positive* and *negative* – hitherto in vogue and partly also to the differing states of the excised eyes observed.

Waller, in his excellent and detailed work on the retinal currents of the frog, has shown how the sign of response is reversed in the moribund condition of the eye.

As to the confusion arising from our present terminology, we must remember that the term *positive* or *negative* is used with regard to a current of reference – the so-called current of injury.

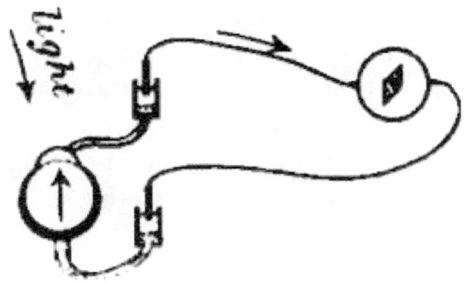

Fig. 95 **Retinal response to light**

The current of response is from the nerve to the retina.

When the two galvanometric contacts are made, one with the cut end of the nerve, and the other on the uninjured cornea, a current of injury is found which *in the eye* is from the nerve to the retina. In the normal freshly excised eye, the current of response due to the action of light on the retina is always from the nerve, which is not directly stimulated by light, to the retina, that is, from the *less* excited to the *more* excited (*Figure 95*). This current of response flows, then, in the same direction as the existing current of reference – the current of injury – and may therefore be called *positive*. Unfortunately the current of injury is very often apt to change its sign; it then flows through the eye from the cornea to the nerve. And now, though the current of response due to light may remain unchanged in direction, still, owing to the reversal of the current of reference, it will appear as *negative*. That is to say, though its absolute direction is the same as before, its relative direction is altered.

I have already advocated the use of the term *positive* for currents which flow towards the stimulated, and *negative* for those whose flow is away from the stimulated. If such a convention be adopted, no confusion can arise, even when, as in the given cases, the currents of injury undergo a change of direction.

## Normal response positive

The normal effect of light on the retina, as noticed by all the observers already mentioned, is a positive variation, during exposure to light of not too long a duration. Cessation of light is followed by recovery. On these points there is general agreement amongst investigators. Deviations are regarded as due to abnormal conditions of the eye, owing to rough usage, or to the rapid approach of death. For just as in the dying plant we found occasional reversals from negative to positive response, so in the dying retina the response may undergo changes from the normal positive to negative.

The sign of response, as we have already seen in numerous cases, depends very much on the molecular condition of the sensitive substance, and if this condition is in any way changed, it is not surprising that the character of the response should also undergo alteration.

Unlike muscle in this, successive retinal responses exhibit little change, for, generally speaking, fatigue is very slight, the retina recovering quickly even under strong light if the exposure is not too long. In exceptional cases, however, fatigue, or its converse, the staircase effect, can be observed.

## Inorganic response under the stimulus of light

It may now be asked whether such a complex vital phenomenon as retinal response could have its counterpart in non-living response. Taking a rod of silver, we can beat out one end into the form of a hollow cup, sensitising the inside by exposing it for a short time to vapour of bromine. The cup is now filled with water, and connection made with a galvanometer by non-polarisable electrodes. There will now be a current due to difference between the inner surface and the rod. This may be balanced, however, by a compensating E.M.F.

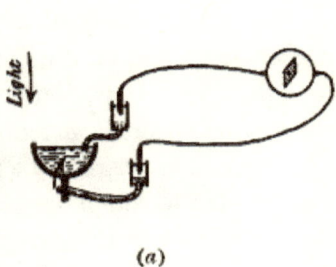 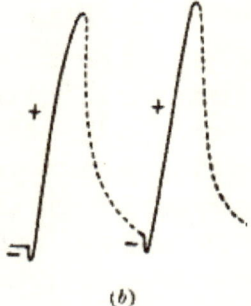

(a)  (b)

### Fig. 96 **Record of responses to light given by the sensitive cell**

Thick lines represent the effect during illumination, dotted lines the recovery in darkness. Note the preliminary negative twitch, which is sometimes also observed in responses of frog's retina.

We have thus an arrangement somewhat resembling the eye, with a sensitive layer corresponding to the retina, and the less sensitive rod corresponding to the conducting nerve stump (*Figure 96a*).

The apparatus is next placed inside a black box, with an aperture at the top. By means of an inclined mirror, light may be thrown down upon the sensitive surface through the opening.

On exposing the sensitive surface to light, the balance is at once disturbed, and a responsive current of positive character produced. The current, that is to say, is from the less to the more stimulated sensitive layer. On the cessation of light, there is a fairly quick recovery (*Figure 96b*).

The character and the intensity of E.M. variation of the sensitive cell depend to some extent on the process of preparation. The particular cell with which most of the following experiments were carried out usually gave rise to a positive variation of about .008 volt when acted on for one minute by the light of an incandescent gas burner which was placed at a distance of 50 cm.

## Typical experiment on the electrical effect induced by light

This subject of the production of an electrical current by the stimulus of light would appear at first sight very complex. But we shall be able to advance naturally to a clear understanding of its most complicated phenomena if we go through a preliminary consideration of an ideally simple case.

We have seen, in our experiments on the mechanical stimulation of, for example, tin, that a difference of electric potential was induced between the more stimulated and less stimulated parts of the same rod, and that a current could thus be obtained, on making suitable electrolytic connections. Whether the more excited was zincoid or cuproid depended on the substance and its molecular condition.

Let us now imagine the metal rod flattened into a plate, and one face stimulated by light, while the other is protected. Would there be a difference of potential induced between the two faces of this same sheet of metal?

Let two blocks of paraffin be taken and a large hole drilled through both. Next, place a sheet of metal between the blocks, and pour melted paraffin round the edge to seal up the junction, the two open ends being also closed by panes of glass. We shall have then two compartments separated by the sheet of metal, and these compartments may be filled with water through the small apertures at the top (*Figure 97a*).

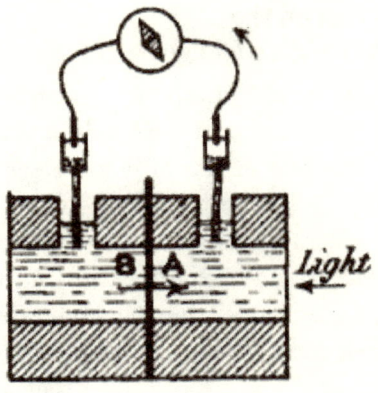

**Fig. 97 (a)**

A, B are the two faces of a brominated sheet of silver. One face, say A, is acted on by light. The current of response is from B to A, across the plate.

**Fig. 97 (b) Record of responses obtained from the above cell**

Ten seconds' exposure to light followed by fifty seconds' recovery in the dark. Thick lines represent action in light, dotted lines represent recovery.

The two liquid masses in the separated chambers thus make perfect electrolytic contacts with the two faces A and B of the sheet of metal. These two faces may be put in connection with a galvanometer by means of two non-polarisable electrodes, whose ends dip into the two chambers. If the sheet of metal has been properly annealed, there will now be no difference of potential between the two faces, and no current in the galvanometer. If the two faces are *not* molecularly similar, however, there will be a current, and the electrical effects to be subsequently described will act additively, in an algebraical sense. Let one face now be exposed to the stimulus of light. A responsive current will be found to flow, from the less to the more stimulated face, in some cases, and in others in an opposite direction.

It appears at first very curious that this difference of electric potential should be maintained between opposite faces of a very thin and highly conducting sheet of metal, the intervening distance between the opposed surfaces being so extremely small, and the electrical resistance quite infinitesimal. A homogeneous sheet of metal has become by the unequal action of light, molecularly speaking, heterogeneous. The two opposed surfaces are thrown into opposite kinds of electric condition, the result of which is as if a certain thickness of the sheet, electrically speaking, were made zinc-like, and the rest copper-like.

From such unfamiliar conceptions, we shall now pass easily to others to which we are more accustomed. Instead of two opposed surfaces, we may obtain a similar response by unequally lighting different portions of the same surface.

Taking a sheet of metal, we may expose one half, say A, to light, the other half, B, being screened. Electrolytic contacts are made by plunging the two limbs in two vessels which are in connection with the two non-polarisable electrodes E and E' (*Figure 98a*). On illumination of A and B alternately, we shall now obtain currents flowing alternately in opposite directions.

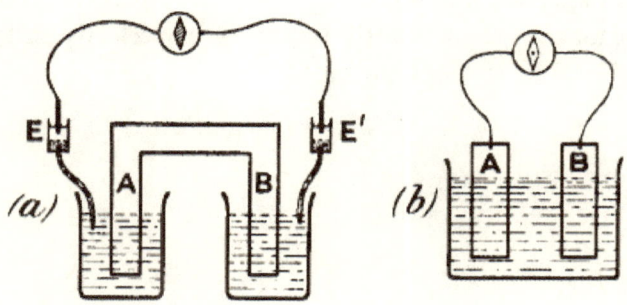

Fig. 98 **Modification of the sensitive cell**

Just as in the strain cells the galvanometer contact was transferred from the electrolytic part to the metallic part of the circuit, so we may next, in an exactly similar manner, cut this plate into two, and connect these directly to the galvanometer, electrolytic connection being made by partially plunging them into a cell containing water. The posterior surfaces of the two half-plates may be covered with a non-conducting coating. And we arrive at a typical photo-electric cell (*Figure 98b*). These considerations will show that the eye is practically a photo-electric cell.

We shall now give detailed experimental results obtained with the sensitive silver-bromide cell, and compare its response curve with those of the retina. A series of uniform light stimuli gives rise to uniform responses, which show very little sign of fatigue. How similar these response curves are to those of the retina will be seen from a pair of records given below, where *Figure 99* shows responses of frog's retina, and *Figure 100* gives the responses obtained with the sensitive silver cell.

It was said that the responses of the retina are uniform. This is only approximately true. In addition to numerous cases of uniform responses, Waller finds instances of 'staircase' increase, and its opposite, slight fatigue. In the record here given of the silver cell, the staircase effect is seen at the beginning, and followed by slight fatigue. I have other records where for a very long time the responses are perfectly uniform, there being no sign of fatigue.

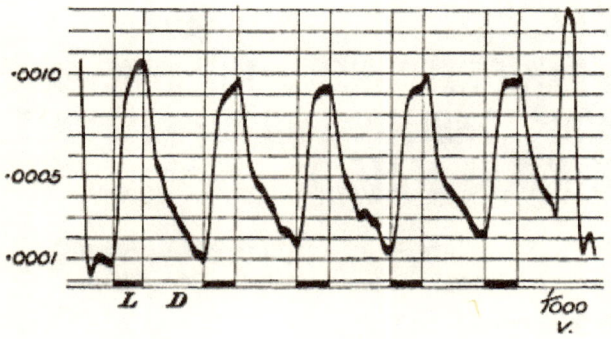

### Fig. 99 **Responses to light in frog's retina**

Illumination L for one minute, recovery in dark for two minutes during obscurity D. (Waller.)

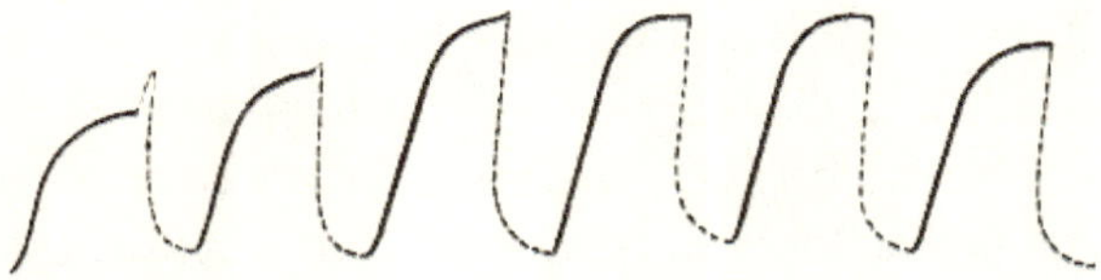

### Fig. 100 **Responses in sensitive silver cell**

Illumination for one minute and obscurity for one minute. Thick line represents record during illumination, dotted line recovery during obscurity.

Another curious phenomenon sometimes observed in the response of retina is an occasional slight increase of response immediately on the cessation of light, after which there is the final recovery. An indication of this is seen in the second and fourth curves in *Figure 99*. Curiously enough, this abnormality is also occasionally met with in the responses of the silver cell, as seen in the first two curves of *Figure 100*. Other instances will be given later.

# Chapter 18
# INORGANIC RESPONSE – INFLUENCE OF VARIOUS CONDITIONS ON THE RESPONSE TO STIMULUS OF LIGHT

*Effect of temperature – Effect of increasing length of exposure – Relation between intensity of light and magnitude of response – After-oscillation – Abnormal effects: (1) preliminary negative twitch; (2) reversal of response; (3) transient positive twitch on cessation of light; (4) decline and reversal – Résumé.*

We shall next proceed to study the effect, on the response of the sensitive cell, of all those conditions which influence the normal response of the retina. We shall then briefly inquire whether even the abnormalities sometimes met with in retinal responses have not their parallel in the responses given by the inorganic.

## Effect of temperature

It has been found that when the temperature is raised above a certain point, retinal response shows rapid diminution. On cooling, however, response reappears, with its original intensity. In the response given by the sensitive cell, the same peculiarity is noticed. I give below (*Figure 101, a*) a set of response curves for 20°C.

These responses, after showing slight fatigue, became fairly constant. On raising the temperature to 50°C response practically disappeared (*Figure 101b*). But on cooling to the first temperature again, it reappeared, with its original if not slightly greater intensity (*Figure 101c*). A curious point is that while in record (*a*), before warming, slight fatigue is observed, in (*c*), after cooling, the reverse, or staircase effect, appears.

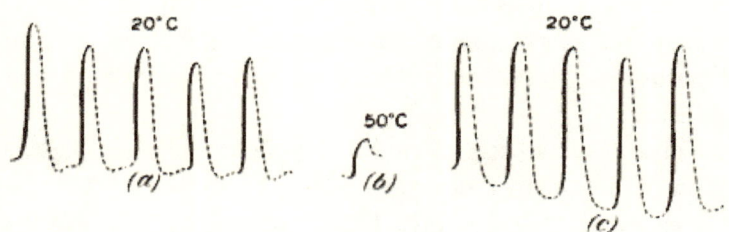

Fig. 101 **Influence of temperature on response**

Illumination 20 seconds, obscurity 40 seconds.

In *(a)* is shown a series of responses at 20°C – the record exhibits slight fatigue. *(b)* is the slight irregular response at 50°C. *(c)* is the record on re-cooling; it exhibits 'staircase' increase.

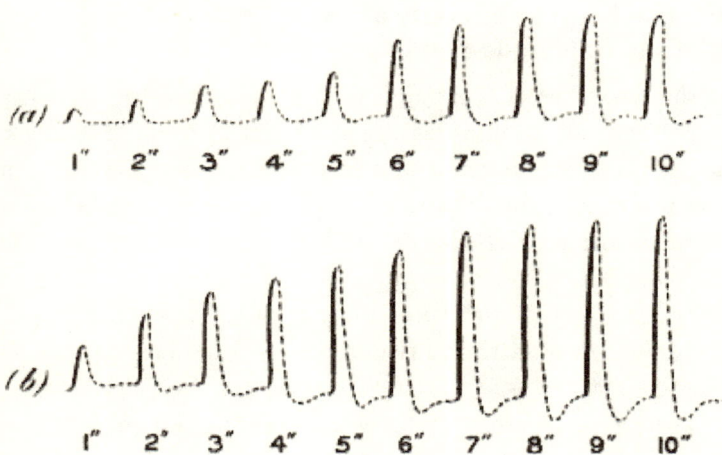

Fig. 102 **Response curves for increasing duration of illumination from 1″ to 10″**

In *(a)* the source of light was at a distance of 50 cm.; in *(b)* it was at a distance of 25 cm. Note the after-oscillation.

## Effect of increasing length of exposure

If the intensity of light is kept constant, the magnitude of response of the sensitive cell increases with the length of exposure. But this soon reaches a limit, after which increase of duration does not increase magnitude of effect. Too long an exposure may however, owing to fatigue, produce an actual decline.

I give here two sets of curves (*Figure 102*) illustrating the effect of lengthening exposure. The intensities of light in the two cases are as 1 to 4. (The incandescent burner was in the two cases at distances 50 and 25 cm respectively.) It will be observed that beyond eight seconds' exposure the responses are approximately uniform. Another noticeable fact is that with long exposure there is an after-oscillation. This growing effect with lengthening exposure and attainment of limit is exactly paralleled by responses of retina under similar conditions.

## Relation between intensity of light and magnitude of response

In the responses of retina, it is found that increasing intensity of light produces an increasing effect. But the rate of increase is not uniform; increase of effect does not keep pace with increase of stimulus. Thus a curve giving the relation between stimulus and response is concave to the axis which represents the stimulus.

The same is true of the sensation of light. That is to say, within wide limits, intensity of sensation does not increase so rapidly as stimulus.

This particular relation between stimulus and effect is also exhibited in a remarkable manner by the sensitive cell. For a constant source of light I used an incandescent burner, and graduated the intensity of the incident light by varying its distance from the sensitive cell. The intensity of light incident on the cell, when the incandescent burner is at a distance of 150 cm., has been taken as the arbitrary unit. In order to make allowance for the possible effects of fatigue I took two successive series of responses (*Figure 103*).

In the first, records were taken with intensities diminishing from 7 to 1, and immediately afterwards increasing from 1 to 7, in the second.

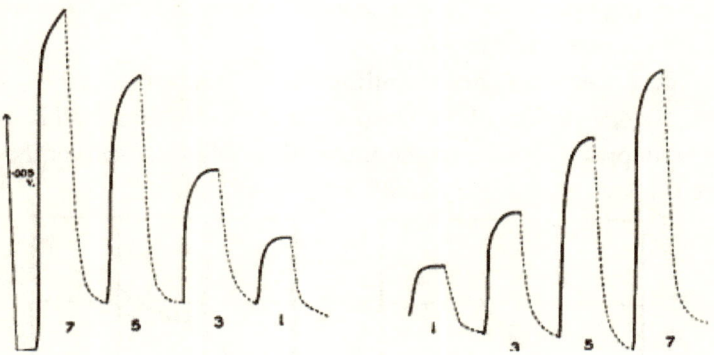

**Fig. 103 Responses of sensitive cell to various intensities of light**

On the left the responses are for diminishing intensities in the ratios of 7,5,3, and 1. On the right they are for the increasing intensities 1, 3, 5, and 7. The thick lines are records during exposures of one minute; the dotted lines represent recoveries for one minute.

## Table giving response to varying intensities of light

(The intensity of an incandescent gas burner at a distance of 150 cm is taken as unit.)

| Intensity of light | response (Light diminishing) | response (Light increasing) | Mean | Value in volts |
|---|---|---|---|---|
| 7 | 43 | 39 | 41 | $63.0 \times 10^{-}$ volt |
| 5 | 31 | 29 | 30 | $46.1 \times 10^{-}$ volt |
| 3 | 18.5 | 17.5 | 18 | $27.7 \times 10^{-}$ volt |
| 1 | 10 | 9 | 9.5 | $14.6 \times 10^{-}$ volt |

As the zero point was slightly shifted during the course of the experiment, the deflection in each curve was measured from a line joining the beginning of the response to the end of its recovery. A mean deflection, corresponding to each intensity, was obtained by taking the average of the descending and ascending readings. The two sets of readings did not, however, vary to any marked extent.

The deflections corresponding to the intensities 1, 3, 5, 7, are, then, as 9.5 to 18, to 30, to 41. If the deflections had been strictly proportionate to the intensities of light stimulus they would have been as 9.5 to 28.5, to 47.5, to 66.5.

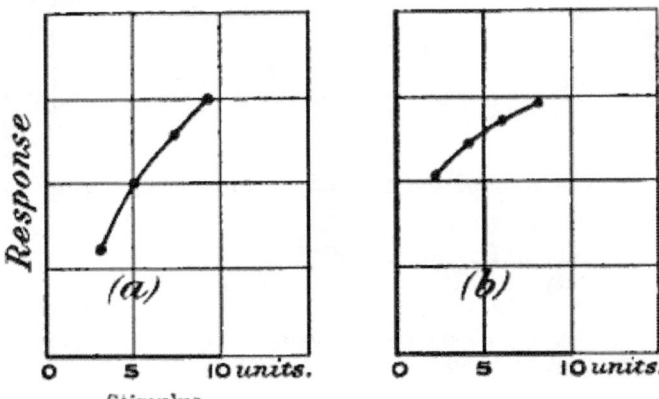

Fig. 104 **Curves giving the relation between intensity of light and magnitude of response**

In *(a)* sensitive cell, *(b)* in frog's retina.

In another set of records, with a different cell, I obtained the deflections of 6, 10, 13, 15, corresponding to light intensities of 3, 5, 7, and 9.

The two curves in *Figure 104*, giving the relation between response and stimulus, show that in the case of inorganic substances, just as in the retina (Waller), magnitude of response does not increase so rapidly as stimulus.

## After-oscillation

When the sensitive surface is subjected to the continued action of light, the E.M. effect attains a maximum at which it remains constant for some time. If the exposure is maintained after this for a longer period, there will be a decline, as we found to be the case in other instances of continued stimulation. The appearance of this decline, and its rapidity, depends on the particular condition of the substance.

When the sensitive element is considerably strained by the action of light, and if that light is cut off, there is a rebound towards recovery and a subsequent after-oscillation. That is to say, the curve of recovery falls below the zero point, and then slowly oscillates back to the position of equilibrium. We have already seen an instance of this in *Figure 102*. Above is given a series of records showing the appearance of decline, from too long-continued exposure and recovery, followed by after-oscillation on the cessation of light (*Figure 105*). Certain visual analogues to this phenomenon will be discussed later.

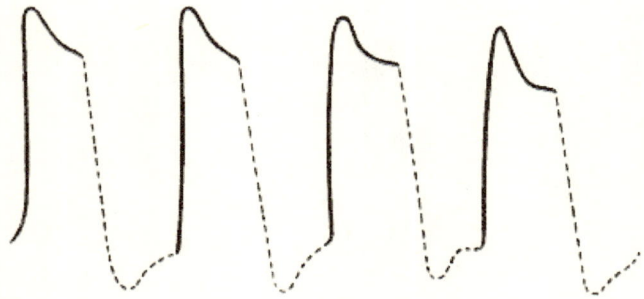

Fig. 105 **After-oscillation**

Exposure for one minute followed by obscurity for one minute. Note the decline during illumination, and after-oscillation in darkness.

## Abnormal effects

We have already discussed all the normal effects of the stimulus of light on the retina, and their counterparts in the sensitive cell. But the retina undergoes molecular changes when injured, stale, or in a dying condition, and under these circumstances various complicated modifications are observed in the response.

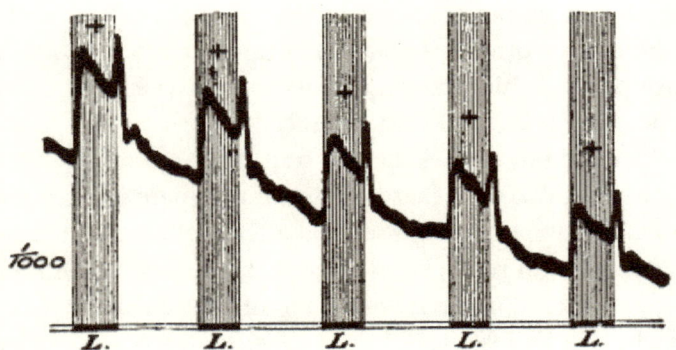

Fig. 106 **Transient positive augmentation given by the frog's retina on the cessation of light 'L'** (Waller)

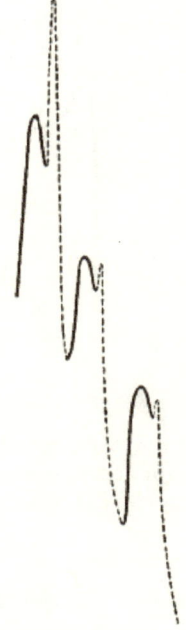

Fig. 107 **Responses in silver cell**

The thick line represents response during light (half a minute's exposure), and the dotted line represents the recovery during darkness. Note the terminal positive twitch.

## 1) Preliminary negative twitch

When the light is incident on the frog's retina, there is sometimes a transitory negative variation, followed by the normal positive response. This is frequently observed in the sensitive cell (see *Figure 96b*).

## 2) Reversal of response

Again, in a stale retina, owing to molecular modification the response is apt to undergo reversal (Waller). That is to say, it now becomes negative. In working with the same sensitive cell on different days I have found it occasionally exhibiting this reversed response.

## 3) Transient rise of current on cessation of light

Another very curious fact observed in the retina by Kuhne and Steiner is that immediately on the stoppage of light there is sometimes a sudden increase in the retinal current, before the usual recovery takes place. This is very well shown in the series of records taken by Waller (*Figure 106*). It will be noticed that on illumination the response curve rises, that continued illumination produces a decline, and that on the cessation of light there is a transient rise of current. I give here a series of records which will show the remarkable similarity between the responses of the cell and retina, in respect even to abnormalities so marked as those described (*Figure 107*).

I may mention here that some of these curious effects, that is to say, the preliminary negative twitch and sudden augmentation of the current on the cessation of light, have also been noticed by Minchin in photo-electric cells.

## 4) Decline and reversal

We have seen that under the continuous action of light, response begins to decline. Sometimes this process is very rapid, and in any case, under continued light, the deflection falls.

(1) The decline may nearly reach zero. If now the light is cut off there is a rebound towards recovery *downwards*, which

carries it below zero, followed by an after-oscillation (*Figure 108, a*).

(2) If the light is continued for a longer time, the decline goes on even below zero; that is to say, the response now becomes apparently negative. If, now, the light is stopped, there is a rebound upwards to recovery, with, generally speaking, a slight preliminary twitch downwards (*Figure 108b,c*). This rebound carries it back, not only to the zero position, but sometimes beyond that position. We have here a parallel to the following observation of Dewar and McKendrick:

> "When diffuse light is allowed to impinge on the eye of the frog, after it has arrived at a tolerably stable condition, the natural E.M.F. is in the first place increased, then diminished; during the continuance of light it is still slowly diminished to a point where it remains tolerably constant, and on the removal of light there is a sudden increase of the E.M. power nearly up to its original position." [18]

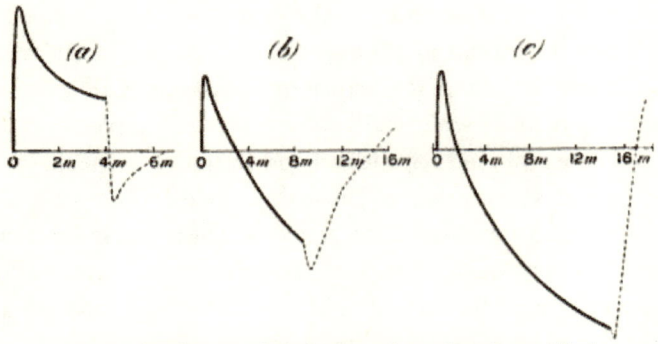

Fig. 108 **Decline under the continued action of light**

a) Decline short of zero; on stoppage of light, rebound

b) downwards to zero; after-oscillation.

c) Decline below zero; on stoppage of light, rebound towards zero, with preliminary negative twitch.

d) The same, decline further down; negative twitch almost disappearing.

(3) I have sometimes obtained the following curious result. On the incidence of light there is a response, say, upward. On the continuation of light the response declines to zero and remains at the zero position, there being no further action during the continuation of stimulus. But on the cessation of light stimulus, there is a response downwards, followed by the usual recovery. This reminds us of a somewhat similar responsive action produced by constant electric current on the muscle. At the moment of 'make' there is a responsive twitch, but afterwards the muscle remains quiescent during the passage of the current, but on breaking the current there is seen a second responsive twitch.

### Resumé

So we see that the response of the sensitive inorganic cell to the stimulus of light is in every way similar to that of the retina. In both we have, under normal conditions, a positive variation; in both, the intensity of response up to a certain limit increases with the duration of illumination; it is affected, in both alike, by temperature; in both there is comparatively little fatigue; the increase of response with intensity of stimulus is similar in both; and finally, even in abnormalities – such as reversal of response, preliminary negative twitch on commencement, and terminal positive twitch on cessation of illumination, and decline and reversal under continued action of light – parallel effects are noticed.

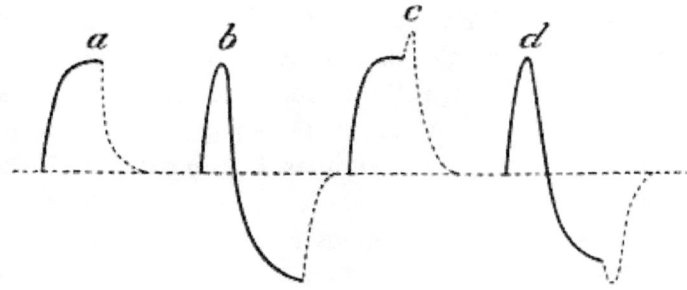

Fig. 109 **Certain after-effects of light**

We may notice here certain curious relations even in these abnormal responses (*Figure 109*). If the equilibrium position remain always constant, then it is easy to understand how, when the rising curve has attained its maximum, on the cessation of light, recovery should proceed *downwards*, towards the equilibrium position (*Figure 109a*). One can also understand how, after reversal by the continued action of light, there should be a recovery *upwards* towards the old equilibrium position (*Figure 109b*). What is curious is that in certain cases we get, on the stoppage of light, a preliminary twitch away from the zero or equilibrium position, upwards as in (*c*) (compare also *Figure 107*) and downwards as in (*d*) (compare also *Figure 108b*).

In making a general summary, finally, of the effects produced by stimulus of light, we find that there is not a single phenomenon in the responses, normal or abnormal, exhibited by the retina which has not its counterpart in the sensitive cell constructed of inorganic material.

### Footnotes

18. *Proc. Roy. Soc. Edin.*, 1873, p.153.

# Chapter 19
# VISUAL ANALOGUES

*Effect of light of short duration  –  After-oscillation
– Positive and negative after-images  –  Binocular
alternation of vision  –  Period of alternation
modified by physical condition  –  After-images and
their revival  –  Unconscious visual impression.*

We have already referred to the electrical theory of the visual impulse. We have seen how a flash of light causes a transitory electric impulse not only in the retina, but also in its inorganic substitute. Light thus produces not only a visual but also an electrical impulse, and it is not improbable that the two may be identical. Again, varying intensities of light give rise to corresponding intensities of current, and the curves which represent the relation between the increasing stimulus and the increasing response have a general agreement with the corresponding curve of visual sensation. In this chapter we shall see how this electrical theory not only explains in a simple manner ordinary visual phenomena, but also concerns others which are more obscure.

We have seen in our silver cell that if the molecular conditions of the anterior and posterior surfaces were exactly similar, there would be no current. In practice, however, this is seldom the case. There is, generally speaking, a slight difference, and a feeble current in the circuit. It is thus seen that there may be an existing feeble current, to which the effect of light is added algebraically. The stimulus of light may thus increase the existing current of darkness (positive variation). On the cessation of light again, the current of response disappears and there remains only the feeble original current.

In the case of the retina, also, it is curious to note that on closing the eye the sensation is not one of absolute darkness, but there is a general feeble sensation of light, known as 'the intrinsic light of the retina.' The effect produced by external light is superposed on this intrinsic light, and certain curious results of this algebraical summation will be noticed later.

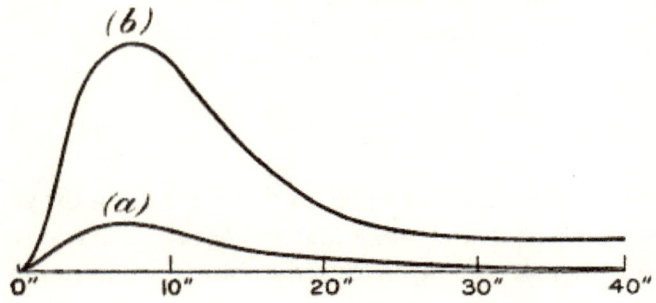

Fig. 110 **Response curves of the sensitive silver cell**

Showing greater persistence of after-effect when the stimulus is strong.

(*a*) Short exposure of 2 seconds to light of intensity 1;
(*b*) short exposure of 2 seconds to light nine times as strong.

## Effect of light of short duration

If we subject the sensitive cell to a flash of radiation, the effect is not instantaneous but grows with time. It attains a maximum some little time after the incidence of light, and the effect then gradually passes away. Again, as we have seen previously with regard to mechanical strain, the after-effect persists for a slightly longer time when the stimulus is stronger. The same is true of the after-effect of the stimulus of light. Two curves which exhibit this are given above (*Figure 110*).

With regard to the first point – that the maximum effect is attained some time after the cessation of a short exposure – a corresponding experiment on the eye may be made as follows:

at the end of a tube is fixed a glass disc coated with lampblack, on which, by scratching with a pin, some words are written in transparent characters. The length of the tube is so adjusted that the disc is at the distance of most distinct vision from the end of the tube applied to the eye. The blackened disc is turned towards a source of strong light, and a short exposure is given by the release of a photographic shutter interposed between the disc and the eye. On closing the eye, immediately after a short exposure, it will at first be found that there is hardly any well-defined visual sensation; after a short time, however, the writing on the blackened disc begins to appear in luminous characters, attains a maximum intensity, and then fades away. In this case the stimulus is of short duration, the light being cut off before the maximum effect is attained. The after-effect here is *positive*, there being no reversal or interval of darkness between the direct image and the after-image, the one being merely the continuation of the other. But we shall see that if light is cut off after a maximum effect is attained by long exposure, that the immediate after-image would be negative (see below).

The relative persistence of after-effect of lights of different intensities may be shown in the following manner:

If a bold design is traced with magnesium powder on a blackened board and fired in a dark room, the observer not being acquainted with the design, the instantaneous flash of light, besides being too quick for detailed observation, is obscured by the accompanying smoke. But if the eyes are closed immediately after the flash, the feebler obscuring sensation of smoke will first disappear, and will leave clear the more persistent after-sensation of the design, which can then be read distinctly. In this manner I have often been able to see distinctly, on closing the eyes, extremely brief phenomena of light which could not otherwise have been observed, owing either to their excessive rapidity or to their dazzling character.[19]

## After-oscillation

In the case of the sensitive silver cell, we have seen (*Figure 105*), when it has been subjected for some time to strong light, that the current of response attains a maximum, and that on the stoppage of the stimulus there is an immediate rebound towards recovery. In this rebound there may be an overshooting of the equilibrium position, and an after-oscillation is thus produced.

If there has been a feeble initial current, this oscillatory after-current, by algebraical summation, will cause the current in the circuit to be alternately weaker and stronger than the initial current.

## Visual recurrence

Translated into the visual circuit, this would mean an alternating series of after-images. On the cessation of light of strong intensity and long duration, the immediate effect would be a negative rebound, unlike the positive after-effect which followed a short exposure.

The next rebound is positive, giving rise to a sensation of brightness. This will go on in a recurrent series.

If we look for some time at a very bright object, preferably with one eye, on closing the eye there is an immediate dark sensation followed by a sensation of light. These go on alternating and give rise to the phenomena of recurrent vision. With the eyes closed, the positive or luminous phases are the more prominent.

This phenomenon may be observed in a somewhat different manner. After staring at a bright light we may look towards a well-lit wall. The dark phases will now become more noticeable. If, however, we look towards a dimly lighted wall, both the dark and bright phases will be noticed alternately.

The negative effect is usually explained as being due to fatigue. That position of the retina affected by light is supposed to be 'tired,' and a negative image to be formed as a result of exhaustion. By this exhaustion is meant either the presence of materials causing fatigue, or the breaking-down of the sensitive element of the tissue, or both of these. In such a case we should expect that this fatigue, with its consequent negative image, would gradually and finally disappear on the restoration of the retina to its normal condition.

We find, however, that this is not the case, for the negative image recurs alternating with the positive. The accepted theory of fatigue is incapable of explaining this phenomenon.

In the sensitive silver cell, we found that the molecular strain produced by light gave rise to a current of response, and that on the cessation of light an oscillatory after-effect was produced. The alternating after-effect in the retina points to an exactly similar process.

## Binocular alternation of vision

It was while experimenting on the phenomena of recurrent vision that I discovered the curious fact that in normal eyes the two do not see equally well at a given instant, but that the visual effect in each eye undergoes fluctuation from moment to moment, in such a way that the sensation in the one is complementary to that in the other, the sum of the two sensations remaining approximately constant. Thus they take up the work of seeing, and then, relatively speaking, resting, alternately. This division of labour, in binocular vision, is of obvious advantage.

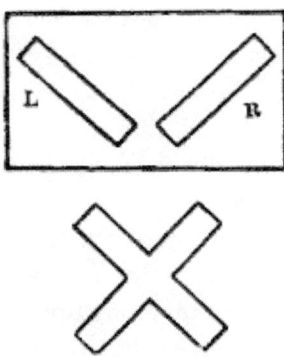

Fig. 111 **Stereoscopic design**

As regards maximum sensation in the two retinæ there is then a relative retardation of half a period. This may be seen by means of a stereoscope, carrying, instead of stereophotographs, incised plates through which we look at light. The design consists of two slanting cuts at a suitable distance from each other. One cut, R, slants to the right, and the other, L, to the left (see *Figure 111*).

When the design is looked at through the stereoscope, the right eye will see, say R, and the left L, the two images will appear superimposed, and we see an inclined cross. When the stereoscope is turned towards the sky, and the cross looked at steadily for some time, it will be found, owing to the alternation already referred to, that while one arm of the cross begins to be dim, the other becomes bright, and *vice versa*. The alternate fluctuations become far more conspicuous when the eyes are closed; the pure oscillatory after-effects are then obtained in a most vivid manner.

After looking through the stereoscope for ten seconds or more, the eyes are closed. The first effect observed is one of darkness, due to the rebound. Then *one* luminous arm of the cross first projects across the dark field, and then slowly disappears, after which the second (perceived by the other eye) shoots out suddenly in a direction athwart the first. This alternation proceeds for a long time, and produces the curious effect of two luminous blades crossing and recrossing each other.

Another method of bringing out the phenomenon of alternation in a still more striking manner is to look at two different sets of writing, with the two eyes. The resultant effect is a blur, due to superposition, and the inscription cannot be read with the eyes open. But on closing them, the composite image is analysed alternately into its component parts, and thus we are enabled to read better with eyes shut than open.

This period of alternation is modified by age and by the condition of the eye. It is, generally speaking, shorter in youth. I have seen it vary in different individuals from one second to ten seconds or more. About four seconds is the most usual. With the same individual, again, the period is somewhat modified by previous conditions of rest or activity. Very early in the morning, after sleep, it is at its shortest. I give below a set of readings given by an observer:

|         | Period    |         | Period      |
|---------|-----------|---------|-------------|
| 8 a.m.  | 3 seconds | 6 p.m.  | 5.4 seconds |
| 12 noon | 4 seconds | 9 p.m.  | 5.6 seconds |
| 3 p.m.  | 5 seconds | 11 p.m. | 6.5 seconds |

Again, if one eye is cooled and the other warmed, the retinal oscillation in one eye is quicker than in the other. The quicker oscillation overtakes the slower, and we obtain the curious phenomenon of 'visual beats'.

## After-images and their revival

In the experiment with the stereoscope and the design of the cross, the after-images of the cross seen with the eyes closed are at first very distinct – so distinct that any unevenness at the edges of the slanting cuts in the design can be distinctly made out. There can thus be no doubt of the 'objective' nature of the strain impression on the retina, which on the cessation of direct stimulus of light gives rise to after-oscillation with the concomitant visual recurrence. This recurrence may therefore be taken as a proof of the physical strain produced on the retina. The recurrent after-image is very distinct at the beginning and becomes fainter at each repetition; a time comes when it is difficult to tell whether the image seen is the objective after-effect due to strain or merely an effect of 'memory'. In fact there is no line of demarcation between the two, one simply merges into the other. That this 'memory' image is due to objective strain is rendered evident by its recurrence.

In connection with this it is interesting to note that some of the undoubted phenomena of memory are also recurrent.

> "Certain sensations for which there is no corresponding process outside the body are generally grouped for convenience under this term [memory]. If the eyes are closed and a picture is called to memory, it will be found that the picture cannot be held, but will repeatedly disappear and appear." [20]

The visual impressions and their recurrence often persist for a very long time. It usually happens that owing to weariness the recurrent images disappear; but in some instances, long after this disappearance, they will spontaneously reappear at most unexpected moments.

In one instance the recurrence was observed in a dream, about three weeks after the original impression was made. In connection with this, the revival of images, on closing the eyes at night, that have been seen during the day is extremely interesting.

## Unconscious visual impression

While repeating certain experiments on recurrent vision, the above phenomenon became prominent in an unexpected manner. I had been intently looking at a particular window, and obtaining the subsequent after-images by closing the eye; my attention was concentrated on the window, and I saw nothing but the window either as a direct or as an after effect. After this had been repeated a number of times, I found on one occasion, after closing the eye, that, owing to weariness of the particular portion of the retina, I could no longer see the after-image of the window; instead of this I however saw distinctly a circular opening closed with glass panes, and I noticed even the jagged edges of a broken pane.

I was not aware of the existence of a circular opening higher up in the wall. The image of this had impressed itself on the retina without my knowledge, and had undoubtedly been producing the recurrent images which remained unnoticed because my principal field of after-vision was filled up and my attention directed towards the recurrent image of the window.

When this failed to appear, my field of after-vision was relatively free from distraction, and I could not help seeing what was unnoticed before.

It thus appears that, in addition to the images impressed in the retina of which we are conscious, there are many others which are imprinted without our knowledge. We fail to notice them because our attention is directed to something else. But at a subsequent period, when the mind is in a passive state, these impressions may suddenly revive owing to the phenomenon of recurrence.

This observation may provide an explanation of some of the phenomena connected with ocular phantoms and hallucinations not traceable to any disease. In these cases the psychical effects produced appear to have no objective cause. Bearing in mind the numerous visual impressions which are being unconsciously made on the retina, it is not at all unlikely that many of these visual phantoms may be due to objective causes.

## Footnotes

19. As an instance of this I mention the experiment which I saw on the quick fusion of metals exhibited at the Royal Institution by Sir William Roberts-Austen (1901), where, owing to the glare and the dense fumes, it was impossible to see what happened in the crucible. But I was able to see every detail on closing the eyes. The effects of the smoke, being of less luminescence, cleared away first, and left the after-image of the molten metal growing clearer on the retina.

20. E. W. Scripture, *The New Psychology*, p. 101.

# Chapter 20
# GENERAL SURVEY AND CONCLUSION

We have seen that stimulus produces a certain excitatory change in living substances, and that the excitation produced sometimes expresses itself in a visible change of form, as seen in muscle. We have also seen that in many other cases, however – as in nerve or retina – there is no visible alteration, but the disturbance produced by the stimulus exhibits itself in certain electrical changes, and that whereas the mechanical mode of response is limited in its application, this electrical form is universal.

This irritability of the tissue, as shown in its capacity for response, either electrical or mechanical, was found to depend on its physiological activity. Under certain conditions it could be converted from a responsive to an irresponsive state, either temporarily as by anæsthetics, or permanently as by poisons. When thus made permanently irresponsive by any means, the tissue was said to have been killed.

We have seen further that from this observed fact – that a tissue when killed passes out of the state of responsiveness into that of irresponsiveness; and from a confusion of 'dead' things with inanimate matter, it has been tacitly assumed that inorganic substances, like dead animal tissues, must necessarily be irresponsive, or incapable of being excited by stimulus – an assumption which has been shown to be gratuitous.

To quote the words of Verworn:

> "This unexplained conception of irritability became the starting-point of *vitalism*, which in its most complete form asserted a dualism of living and lifeless Nature....
>
> The vitalists soon laid aside, more or less completely, mechanical and chemical explanations of vital phenomena, and introduced, as an explanatory principle, an all-controlling, unknown and inscrutable 'force hypermécanique'. While chemical and physical forces are responsible for all phenomena in lifeless bodies, in living organisms this special force induces and rules all vital actions.
>
> Later vitalists, however, attempted no analysis of vital force; they employed it in a wholly mystical form as a convenient explanation of all sorts of vital phenomena....
>
> In place of a real explanation a simple phrase such as 'vital force' was satisfactory, and signified a mystical force belonging to organisms only. Thus it was easy to 'explain' the most complex vital phenomena." [21]

From this position, with its assumption of the super-physical character of response, it is clear that with the discovery of similar effects amongst inorganic substances, the necessity of theoretically maintaining such dualism in Nature must immediately fall to the ground.

In the previous chapters I have shown that not the fact of response alone, but *all* those modifications in response which occur under various conditions take place in plants and metals just as they do in animal tissues. It may now be useful to make a general survey of these phenomena, as exhibited in the three classes of substances.

We have seen that the wave of molecular disturbance in a living animal tissue under stimulus is accompanied by a wave of electrical disturbance; that in certain types of tissue the stimulated is relatively positive to the less disturbed, while in others it is the reverse; that it is essential to the obtaining of electric response to have the contacts leading to the galvanometer unequally affected by excitation; and finally that this is accomplished either:

(1) by 'injuring' one contact, so that the excitation produced there would be relatively feeble, or

(2) by introducing a perfect block between the two contacts, so that the excitation reaches one and not the other.

Further, it has been shown that this characteristic of exhibiting electrical response under stimulus is not confined to animal, but extends also to vegetable tissues. In these the same electrical variations as in nerve and muscle were obtained by using the method of injury, or that of the block.

In dealing with inorganic substances, and using similar experimental arrangements, we have found the same electrical responses evoked in metals under stimulus.

## Negative variation

In all cases, animal, vegetable, and metal, we can obtain response by the method of negative variation, so called, by reducing the excitability of one contact by physical or chemical means. Stimulus causes a transient diminution of the existing current, the variation depending on the intensity of the stimulus (*Figures 4, 7, 54*).

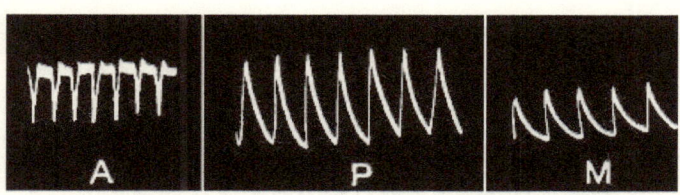

Fig. 112 **Uniform responses in (A) nerve, (P) plant, and (M) metal**

> The normal response in nerve is represented 'down.' In this and following figures, **A** is the record of responses in animal, **P** in plant, and **M** in metal.

## Relation between stimulus and response

In all three classes we have found that the intensity of response increases with increasing stimulus. At very high intensities of stimulus, however, there is a tendency of the response to reach a limit (*Figures 30, 32, 84*). The law that is known as Weber-Fechner's shows a similar characteristic in the relation between stimulus and sensation. And if sensation is a measure of physiological effect we can understand this correspondence of the physiological and sensation curves. We now see further that the physiological effects themselves are ultimately reducible to simple physical phenomena.

## Effects of superposition

In all three types, ineffective stimuli become effective by superposition.

Again, rapidly succeeding stimuli produce a maximum effect, kept balanced by a force of restitution, and continuation of stimulus produces no further effect, in the three cases alike (*Figures 17, 18, 86*).

## Uniform responses

In the responses of animal, vegetable, and metal alike we meet with a type where the responses are uniform (*Figure 112*).

## Fatigue

There is, again, another type where fatigue is exhibited.

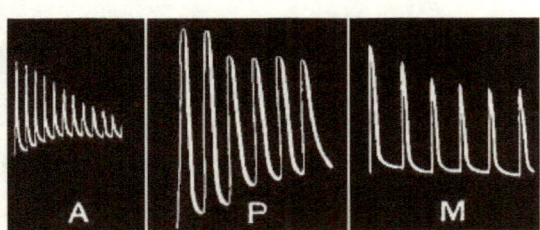

Fig. 113 **Fatigue (A) in muscle, (P) in plant, (M) in metal**

The explanation hitherto given of fatigue in animal tissues – that it is due to dissimilation or breakdown of tissue, complicated by the presence of fatigue-products, while recovery is due to assimilation, for which material is brought by the blood-supply – has long been seen to be inadequate, since the restorative effect succeeds a short period of rest even in excised bloodless muscle. But that the phenomena of fatigue and recovery were not primarily dependent on dissimilation or assimilation becomes self-evident when we find exactly similar effects produced not only in plants, but also in metals (*Figure 113*). It has been shown, on the other hand, that these effects are primarily due to cumulative residual strains, and that a brief period of rest, by removing the overstrain, removes also the sign of fatigue.

## Staircase effect

The theory of dissimilation due to stimulus reducing the functional activity below par, and thus causing fatigue, is directly contradicted by what is known as the 'staircase' effect, where successive equal stimuli produce increasing response. We saw an exactly similar phenomenon in plants and metals, where successive responses to equal stimuli exhibited an increase, apparently through a gradual removal of molecular sluggishness (*Figure 114*).

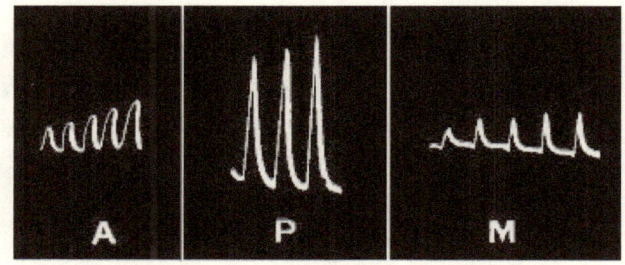

Fig. 114 **'Staircase' in muscle, plant, and metal**

## Increased response after continuous stimulation

An effect somewhat similar, that is to say, an increased response, due to increased molecular mobility, is also shown sometimes after continuous stimulation, not only in animal tissues, but also in metals (*Figure 115*).

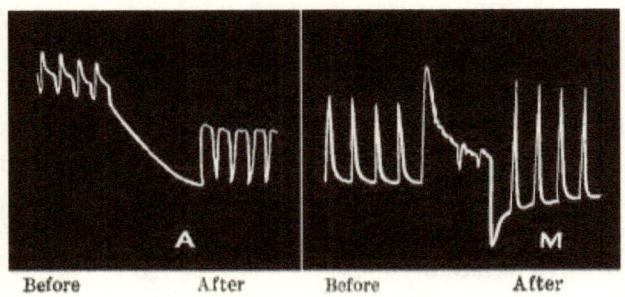

Before    After    Before    After

**Fig. 115 Increased response after continuous stimulation in nerve and metal**

The normal response in animal tissue is represented as 'down,' and in metal 'up.'

## Modified response

In the case of nerve we saw that the normal response, which is negative, sometimes becomes reversed in sign, i.e. positive, when the specimen is stale. In retina again the normal positive response is converted into negative under the same conditions. Similarly, we found that a plant when withering often shows a positive instead of the usual negative response (*Figure 28*). On nearing the death-point, also by subjection to extremes of temperature, the same reversal of response is occasionally observed in plants. This reversal of response due to peculiar molecular modification was also seen in metals.

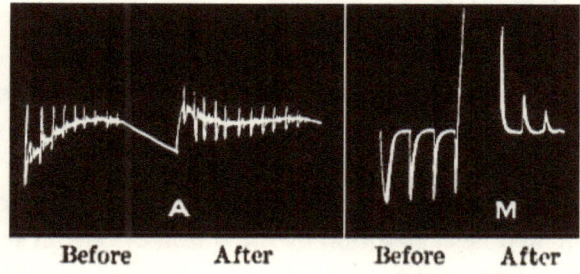

**Fig. 116 Modified abnormal response in (A) nerve and (M) metal converted into normal, after continuous stimulation**

> A is the record for nerve (recording galvanometer not being dead-beat shows after-oscillation); the abnormal 'up' is converted into normal 'down' after continuous stimulation. M is the record for metal, the abnormal 'down' being converted into normal 'up' after like stimulation.

But these modified responses usually become normal when the specimen is subjected to stimulation either strong or long continued (*Figure 116*).

## Diphasic variation

A diphasic variation is observed in nerve if the wave of molecular disturbance does not reach the two contacts at the same moment, or if the rate of excitation is not the same at the two points. A similar diphasic variation is also observed in the responses of plants and metals (*Figures 26, 68*).

## Effect of temperature

In animal tissues response becomes feeble at low temperatures. At an optimum temperature it reaches its greatest amplitude, and, again, beyond a maximum temperature it is very much reduced.

We have observed the same phenomena in plants. In metals too, at high temperatures, the response is very much diminished (*Figures 38, 65*).

## Effect of chemical reagents

Finally, just as the response of animal tissue is exalted by stimulants, lowered by depressants, and abolished by poisons, so also we have found the response in plants and metals undergoing similar exaltation, depression, or abolition.

We have seen that the criterion by which vital response is differentiated is its abolition by the action of certain reagents – the so-called poisons. We find, however, that 'poisons' also abolish the responses in plants and metals (*Figure 117*). Just as animal tissues pass from a state of responsiveness while living to a state of irresponsiveness when killed by poisons, so also we find metals transformed from a responsive to an irresponsive condition by the action of similar 'poisonous' reagents.

The parallel is the more striking since it has long been known with regard to animal tissues that the same drug, administered in large or small doses, might have opposite effects, and in preceding chapters we have seen that the same statement holds good with respect to plants and metals also.

## Stimulus of light

Even the responses of such a highly specialised organ as the retina are strictly paralleled by inorganic responses. We have seen how the stimulus of light evokes in an artificial retina responses which coincide in all their detail with those produced in a real retina.

This was seen in ineffective stimuli becoming effective after repetition, in the relation between stimulus and response, and in the effects produced by temperature; also in the phenomenon of after-oscillation. These similarities went even further, the very abnormalities of retinal response finding their reflection in the inorganic.

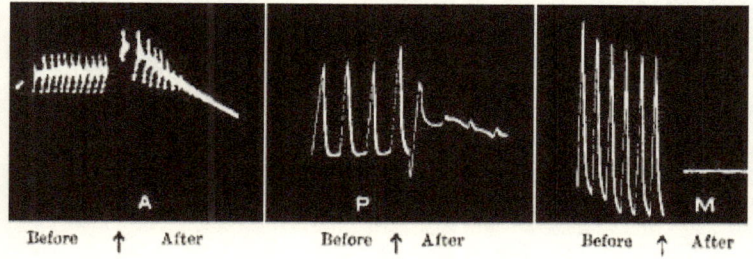

### Fig. 117 **Abolition of response in nerve, plant, and metal by the action of the same 'poison'**

The first half in each set shows the normal response, the second half the abolition of response after the application of the reagent.

Thus living response in all its diverse manifestations is found to be only a repetition of responses seen in the inorganic. There is in it no element of mystery or caprice, such as we must admit to be applied in the assumption of a hypermechanical vital force, acting in contradiction or defiance of those physical laws that govern the world of matter. Nowhere in the entire range of these response-phenomena – inclusive as that is of metals, plants, and animals – do we detect any breach of continuity.

In the study of processes apparently so complex as those of irritability, we must, of course, expect to be confronted with many difficulties. But if these are to be overcome, they, like others, must be faced, and their investigation patiently pursued, without the postulation of special forces whose convenient property it is to meet all emergencies in virtue of their vagueness. If, at least, we are ever to understand the intricate mechanism of the animal machine, it will be granted that we must cease to evade the problems it presents by the use of mere phrases which really explain nothing.

We have seen that amongst the phenomena of response, there is no necessity for the assumption of vital force. They are, on the contrary, physico-chemical phenomena, susceptible to a physical inquiry as definite as any other in inorganic regions.

Physiologists have taught us to read in the response curves a history of the influence of various external agencies and conditions on the phenomenon of life. By these means we are able to trace the gradual diminution of responsiveness by fatigue, by extremes of heat and cold, its exaltation by stimulants, and the arrest of the life process by poison.

The investigations which have just been described may possibly carry us one step further, proving to us that these things are determined, not by the play of an unknowable and arbitrary vital force, but by the workings of laws that know no change, acting equally and uniformly throughout both the organic and the inorganic worlds.

## Footnotes

21. Verworn, *General Physiology*, p. 18.

## ALSO FROM A DISTANT MIRROR

*The Blood and its Third Element*
Antoine Bechamp

*Bechamp or Pasteur?*
Ethel Hume

*Reconstruction by Way of the Soil*
Guy Wrench

*The Soil and Health*
Albert Howard

*The Wheel of Health*
Guy Wrench

*The Soul of the Ape & My Friends the Baboons*
Eugene Marais

*The Soul of the White Ant*
Eugene Marais

*Earthworm*
George Oliver

*My Inventions*
Nikola Tesla

*The Problem of Increasing Human Energy*
Nikola Tesla

*Response in the Living and Non-living*
Jagadish Bose

WWW.ADISTANTMIRROR.COM.AU

# Béchamp or Pasteur?

*A Lost Chapter in the History of Biology*

by Ethel D. Hume

352 pages

This volume contains new editions of two titles.

*Available in paperback, epub and Kindle formats.*

R. Pearson's *Pasteur: Plagiarist, Imposter* was originally published in 1942, and is a succinct introduction to both Louis Pasteur and Antoine Bechamp, and the reasons behind the troubled relationship that they shared for their entire working lives.

Whereas Pearson's work is a valuable introduction to an often complex topic, it is Ethel Douglas Hume's expansive and well-documented *Bechamp or Pasteur? A Lost Chapter in the History of Biology* which provides the main body of evidence. It covers the main points of contention between Bechamp and Pasteur in depth sufficient to satisfy any degree of scientific or historical scrutiny, and it contains, wherever possible, detailed references to the source material and supporting evidence.

Virtually no claim in Ms Hume's book is undocumented. The reader will soon discern that neither Mr Pearson nor Ms Hume could ever be called fans of Pasteur or his 'science'. They both declare their intentions openly; that they wish to contribute to the undoing of a massive medical and scientific fraud.

The text of both titles has been extensively re-edited so as to modernise the use of English, and make the book easier to read than has been the case with previous facsimile editions.

Included are new renderings of all the diagrams that were included in the original edition of *Pasteur: Plagiarist, Imposter*, plus there is a collection of what photographs of Professor Bechamp are available.

# A DISTANT MIRROR

www.ingramcontent.com/pod-product-compliance
Lightning Source LLC
Chambersburg PA
CBHW020903180526
45163CB00007B/2603